Reliable and Sustainable Electric Power and Energy Systems Management

Series editors

Ajit Kumar Verma, Stord, Norway
Rajesh Karki, Saskatoon, Canada
Jaeseok Choi, Jinju, GN, South Korea

More information about this series at http://www.springer.com/series/10855

Nava Raj Karki · Rajesh Karki
Ajit Kumar Verma · Jaeseok Choi
Editors

Sustainable Power Systems

Modelling, Simulation and Analysis

Springer

Editors
Nava Raj Karki
Tribhuvan University
Kathmandu
Nepal

Rajesh Karki
University of Saskatchewan
Saskatoon, SK
Canada

Ajit Kumar Verma
Stord/Haugesund University College
Stord
Norway

Jaeseok Choi
Gyeongsang National University
Jinju, Gyeongsangnam-do
South Korea

ISSN 2510-2524 ISSN 2510-2532 (electronic)
Reliable and Sustainable Electric Power and Energy Systems Management
ISBN 978-981-10-2229-6 ISBN 978-981-10-2230-2 (eBook)
DOI 10.1007/978-981-10-2230-2

Library of Congress Control Number: 2016947778

Printed on acid-free paper

This Springer imprint is published by Springer Nature
The registered company is Springer Science+Business Media Singapore Pte Ltd.

Contents

About the Editors

Nava Raj Karki obtained his M.Sc. (Hons) in Electrical Engineering from Belarussian National Technical University, Minsk in 1995 and Ph.D. in Electrical Engineering from Indian Institute of Technology, Bombay in 2011. He joined Department of Electrical Engineering, Institute of Engineering, Tribhuvan University in 1995 and is currently an Associate Professor and Head of the department. He is actively involved in IEEE activities and is the founding Vice Chair of IEEE Nepal Subsection. Dr. Nava Raj Karki is very active in promoting Academia-Industry collaboration at the Institute of Engineering and various multilateral initiatives like Erasmus Mundus Programme. He has published several papers on power system planning, reliability, distributed generation, and restructuring, and also published a book 'Electrical Engineering Materials' (Alpha Science International Ltd., 2007).

Rajesh Karki obtained his B.E. degree in electrical engineering from the Regional Engineering College (since renamed National Institute of Technology), Durgapur, West Bengal, India in 1991, and M.Sc. and Ph.D. degrees in electrical power engineering in 1997 and 2000, respectively, from the University of Saskatchewan, Canada. He worked for academic institutions and different industries in Nepal between 1991 and 1995, and for GE Industrial Systems, Peterborough, ON, Canada during 2000–2002. He joined the University of Saskatchewan, Canada as an Assistant Professor in 2002, where he currently works as a Professor. During 2005–2012, he chaired the Power Systems Research Group at the University of Saskatchewan. He has served in various capacities in conferences and workshops, given lectures in Canada and has published over 80 papers in reputed international journals and peer reviewed conferences. Dr. Karki has completed several consulting projects on system planning and reliability for Canadian electric utilities. He is a Senior Member of the IEEE, and a Professional Engineer in the Province of Saskatchewan, Canada. His research interests include power system reliability and planning, and reliability modeling and analysis of renewable energy systems.

Ajit Kumar Verma obtained his B.Tech. (Electrical Engineering) and Ph.D. (Reliability Engineering) from Indian Institute of Technology Kharagpur, India in 1984 and 1988 respectively. He joined Indian Institute of Technology Bombay in 1988 where he holds the position of Professor of Electrical Engineering. He is currently working as a Professor of Technical Safety at University College Stord/Haugesund, Norway. He has supervised/co-supervised 28 doctoral and 88 masters thesis in the area of Software Reliability, Reliable Computing, Power Systems Reliability (PSR), Reliability Centered Maintenance (RCM), and Probabilistic Safety/Risk Assessment (PSA). He has over 195 publications in various journals and conferences. He is a senior member of IEEE and a life fellow of IETE. He is the Editor-in-Chief of the *International Journal of Systems Assurance Engineering and Management* (IJSAEM) published by Springer, as well as a member of the Advisory Committee for the journal OPSEARCH also published by Springer. He has served as a Guest Editor of many international journals including *IEEE Transactions on Reliability* for the March 2010 issue. He has been the Conference Chairman of various international conferences and symposia.

Jaeseok Choi is a Professor at Gyeongsang National University, Korea and has been teaching at the university since 1991. He did his masters and obtained his Ph. D. from Korea University in 1984 and 1990, respectively. He completed his postdoctoral research at University of Saskatchewan, Canada in 1996. He has published many papers in reputed international journals and conference proceeding, and undertaken many projects funded by the Korean Electric Power Corporation. His research interests include fuzzy applications, probabilistic production cost simulation, reliability evaluation and outage cost assessment of power systems, and development of flexible methods for power system. His recent research is specially focused on development and extension of methodology at load points of Composite Power Systems.

Introduction

The importance of electrical power and energy in modern society is well established and it is impossible to create a modern society without existence of not only reliable but sustainable power and energy systems. The continued development of systems that are both reliable and sustainable has been the main focus of power system engineers as well as advocates of sustainable development. The challenges and difficulties faced by large-scale power generation and transmission systems are encouraging the development planners to explore and expedite the inclusion of more power supply options regarded as reliable as well as sustainable. There is growing global environmental concern regarding increasing emissions from the electric power generation required to meet rising energy needs and support sustainable and inclusive development. As a result, the integration of renewable sources in electric grids has helped increase the penetration of electricity in many developing and least developed countries, and helped reduce emissions in advanced countries. The advancement of electric vehicles has achieved significant growth in recent years and has the potential of changing the planning and operating paradigm of the power system. The increasing uncertainty associated with planning and operating modern power systems creates significant challenges to maintaining the system efficiency, energy security, and reliability. Modernizing power systems with the application of smart grid technologies is perceived to be a valuable approach to achieve reliability and efficiency with environmental compliance and also help improve energy independence and security. The appropriate use of distributed energy resources in conjunction with state-of-art technology is an important asset in mitigating these problems, and requires considerable study and research. This volume intends to bring out the work of researchers from academia and industry in understanding, modeling, and analysis of the risks associated with implementing sustainable energy technology in power systems. The topics in this volume include the Reliability Assessment of Power Grid Supporting Sustainable Transportation, Hybrid AC/DC Micro-Grids: Solution for High Efficient Future Power Systems, Low Voltage Ride Through of Wind Energy Systems, Reliability Evaluation of Distribution System with Network Reconfiguration and Distributed Generations, Tracing the Unreliability and Recognizing the Weak Parts of a Power System,

Sustainable Energy Optimization in a Smart Microgrid, Renewable Energy Generation System Expansion Planning, Microhydro Based Mini Grid for Sustainable Development of Rural Communities: A Case Study of Nepal.

Chapter 1 titled "Reliability Assessment of Power Grid Supporting Sustainable Transportation" presents analytical models to evaluate the reliability of a power system with PHEV penetration considering the important PHEV characteristics and charging scenarios. The parameters of the PHEV-30 and the transportation data from NHTS 2009 were used to create the relevant models to obtain the load model of a PHEV. A MCS method was applied to combine individual PHEV models and build the overall model of PHEV fleet. The impact of PHEV penetration in a power system was assessed by using the IEEE-RTS as a test system. Different PHEV penetrations and charging scenarios were examined to analyze the results. The results show that the system reliability degrades with PHEV penetration increases, and the negative impact is increased with fast charging which requires a higher power demand for a shorter duration. The negative reliability impacts can, however, be mitigated by shifting and controlling the charging time, which is the area of ongoing research.

Chapter 2 titled "Hybrid AC/DC Micro-Grids: Solution for High Efficient Future Power Systems" presents a new configuration for future power systems which is the hybrid AC/DC gird for high efficient connection of the inherent AC and DC sources and loads. Three phase AC power systems have been in dominant position for over hundred years due to invention of transformer and the inherent characteristic from fossil energy driven rotating machines. However, the gradual changes of load types and distributed renewable generation (DRG) in AC local distribution systems provide food for consideration of adding DC networks. Renewable sources such as fuel cells and solar photovoltaics are DC inherent and should be connected to AC grid through DC/AC conversion techniques, whereas some AC inherent renewable sources like wind generators also need DC links in their conversion systems to increase efficiency and mitigate power variation caused by intermittency and uncertainty. The disadvantage of AC grids for connection of DC inherent sources and loads as well as AC loads with DC links is that additional DC/AC or AC/DC converters are required, which may result in efficiency loss from the reverse conversion. In the other hand DC grids are resurging due to the development and deployment of renewable DC power sources and their inherent advantage for DC loads in commercial, industrial, and residential applications. The number of power conversions in a DC microgrid has been significantly reduced to enhance system energy efficiency. A more likely scenario is the coexistence of both AC and DC microgrids, which is so-called the hybrid AC/DC microgrid in order to reduce processes of multiple reverse conversions in an individual AC or DC microgrid and facilitate the connection of various renewable AC/DC sources and loads to power system. Therefore the concept of hybrid microgrids, which can harmonize both AC and DC sources and loads, has been proposed for future high efficient power systems. Conventional AC and DC grids are interconnected together through the bidirectional AC/DC converter. The component model has been introduced. The control and operation of individual sources and energy storages are presented. The coordination control and power sharing techniques are also introduced.

Chapter 3 titled "Low Voltage Ride Through of Wind Energy Systems" focuses on the importance of low voltage ride through (LVRT) of wind turbines. The investigation of some selected grid codes has revealed that LVRT has different requirements in different grid codes. The results of the investigation have found that higher penetration of wind energy in grid results in stricter LVRT requirements. On the other hand, increasing the wind energy penetration in grid is an ongoing process, so it will be mandatory to increase the LVRT performance of existing wind turbines as well as newly installed units. This chapter has presented some key power circuit topology-based and control strategy-based LVRT improvement methods. While power circuit topology-based strategies are usually more efficient than control strategy-based methods, they need extra power electronic-based electrical circuits which might be expensive or bulky and they may affect the normal operation of wind turbines too. This chapter has also discussed the problems of resynchronization of wind turbines. It is shown that without sufficient LVRT capability, wind turbines would be disconnected from grid during faults and resynchronization of them would be a time-consuming process which may impose large transients to the grid. A resynchronization scheme for automatic reconnection of wind turbines has been presented and also validated through simulation.

Chapter 4 titled "Reliability Evaluation of Distribution System with Network Reconfiguration and Distributed Generations" proposes reliability and power loss evaluation of distribution system using network reconfiguration tool and optimal DG siting and sizing. The objectives are reliability improvement and power loss reduction subjected to various operating constraints. The renewable DG technologies including the synchronous machine-based biomass generator are considered in distribution system expansion planning. The network reconfiguration with DGs and tie-switch placement for the reliability improvement and loss minimization is proposed. Tie-switches are placed at terminal nodes with geographical constraints and their all possible binary combinations are considered. DGs sizing for the predefined DGs are calculated using integer programming for objective, viz., reliability improvement and minimization of losses with penetration level taken for less than total load. Power loss is calculated using forward/backward sweep distribution load flow algorithm in all the cases analyzed. A search-based reconfiguration algorithm has been formulated for finding the optimal switch configuration for the radial distribution system with DG. The optimal DG placement and DG sizing for combined objectives of reliability improvement and loss minimization, using fuzzy logic approach is proposed in two steps. Out of four DG variables, type and number are taken as fixed variables and size and location are considered as variables in the formulation. In the first step, for all DGs combination, their optimal sizes are calculated using nonlinear optimization tool with the improved reliability and real power loss reduction objectives. SAIDI and SAIFI indices are used for the reliability evaluation in this case. In the formulation three indices, power loss, SAIDI and SAIFI are used with weighing factors and taken as combined objective. A combined network reconfiguration and optimal DG siting and sizing is carried out for the objectives, viz., real power loss reduction and reliability improvement for various penetration levels and various loading conditions. It is observed that optimal sitting and sizing of the DGs improve the system performance in terms of real power loss reduction and improvement in the system reliability. Different objectives give different optimal locations and sizes but based on the physical constraints, the final decisions are to be made. The present work may be a guideline to the planners and policymakers for planning the DGs in the distribution systems.

Chapter 5 titled "Tracing the Unreliability and Recognizing the Weak Parts of a Power System" presents two Unreliability tracing principles, i.e., failure component sharing principle (FCSP) and proportional sharing principle (PSP) to derive the UT model and algorithm for a power system. The proposed technique provides a new idea for evaluating the impact of components on the system unreliability and can be used to distribute the unreliability indices of a power system to the system components impartially and reasonably.The models and algorithms for tracing the unreliability are determined using the PSP. The UT indices for individual component are created to recognize the major unreliability contributions (MUC) of the power system. The method provides the system planner/operator with information on how much an individual component contributes to the overall system risk and the system MUC. The basic principles and methods proposed in this paper are general and therefore applicable to both HL-I and HL-II. The illustration application is at HL-I and the RBTS and IEEE-RTS were examined using the proposed method. The results show that the proposed approach has good practicability and has promising prospects in other engineering applications. Power system applications utilizing the proposed UT technique could address using decision analysis to recognize weak system elements, improve the reliability, analyze reliability cost/benefits, and optimize scheduled maintenance. In addition, the proposed UT technique can be easily implemented on the existing reliability evaluation programs with a relatively little additional computation requirement, and conveniently expanded to AC/DC systems, distribution systems or interconnected systems.

Chapter 6 titled "Sustainable Energy Optimization in a Smart Microgrid" presented the methodology for developing a smart microgrid optimization model that can be used to determine the most financially economical combination of microgrid technologies with acceptable level of system reliability. The smart microgrid optimization model assessed various combinations of PV and wind turbine renewable generation in increments of 5 kW varying between 0 and 200 kW. Battery energy storage was also considered in combination with renewable generation in increments of 5 kWh from 0 to 200 kWh. The smart microgrid optimization model was developed in the MATLAB environment and incorporated a smart microgrid management system to achieve additional fuel savings and increased system reliability by utilizing load shedding and load deferral techniques. Microgrid power system reliability was assessed using an SMCS simulation model. The accuracy of the model was verified using the SIPSREL program developed at the University of Saskatchewan. Results indicated that the implementation of renewable generation, battery energy storage, and DSM techniques can substantially reduce the lifetime operational costs of a microgrid while increasing power system reliability. Implementation of energy storage technologies to reduce fossil fuel consumption as well as to increase renewable generation penetration is an area of ongoing research, as is the application of microgrid technology to improve remote power system reliability.

Chapter 7 titled "Renewable Energy Generation System Expansion Planning" proposes a new approach using linear programming to solve the long-term generation mix considering renewable energy resources generators namely; Wind Turbine Generators (WTGs) and Solar Cell Generator (SCGs) with multi-criteria considering CO_2, SO_x, and NO_x emission constraints under the uncertain circumstances. The chapter focuses on the development of tools to analyze the effect of WTGs and SCGs on the best generation mix (BGM) using fuzzy set theory. The method accommodates sensitivity analysis of capacity factors of WTG and SCG. The effectiveness of the proposed approach is demonstrated by applying to solve the flexible generation mix problem on the Korean power system for a planning horizon up to the year 2030. The test system contains nuclear, coal, LNG, oil, pumped-storage hydroplant, WTG and SCG in multi-years period. The economic credits of renewable energy resource in best generation mix problem were proposed in the viewpoint of CF and CER of WTG and SCG. It is noted that the WTG and SCG are competitive generator types in the mix and the WTG and SCG can win the conventional generators mix from for various percentage mix of WTGs and SCGs in the Korean power system. The proposed model and method are useful tools for various economics evaluation in BGM considering renewable energy generators and emission constraints.

Chapter 8 titled "Microhydro-Based Mini grid for Sustainable Development of Rural Communities: A Case Study of Nepal" presents the overview of Microhydro Based Mini Grid for Sustainable Development of Rural Communities in Nepal. Development of energy sector in Nepal demands optimum tapping of Hydropower of different scales; be it through large, medium, or small size hydroplants. Large power plants are often required big capital investment and involve long constructional time, whereas small and microhydro plants (SMHP) can be installed in short time period with local initiative. Since very beginning SMHP are proved to be as cost-effective, indigenous, and reliable source of electrical power. However, they also have been facing hurdles in full-scale development. The main issues are low load factor, low capacity to run big machines, poor end-use pattern, and unavailability of water throughout the year. Besides the expansion of national grid to the MHP vicinity has caused to extinct, thereby wasting the productive asset. This scenario has forced to develop integrated hydropower development model in country, in which MHP can be operated even after the reach of grid. By this, additional revenue will be generated from surplus electricity. Second, parallel operation of two or more MHPs can realized for sharing the common loads. Formation of such MGs would have number of benefits; it would maximize the use of spilling resource, increase load factor and improve the reliability of the overall system. These MGs can be connected with national grid if it is nearby, and operated autonomously in islanded mode and interconnected mode in specified manner. From utility perspective, the MG functions as single entity, which can provide a better platform to interconnect SMHPs to grid? Thereby they can be operated even after the reach of grid in their sites, making MG a sustainable scheme for SMHP development. There are several sites in Nepal with potential of MG. AEPC has reported 14 sites across the country with capacity ranging 100 kW to 1 MW. AEPC also initiated to pilot a MG project in Baglung District by connecting the MHPs along the corridor of Kalung Khola river, which was completed and commissioned in 2011. Presently, It has been drawing attention to the researchers, entrepreneurs, and professionals across the globe for the real-time study of MG. Its success stories have paved the ways for the further development of other MGs in the different parts of country.

Chapter 1
Reliability Assessment of Power Grid Supporting Sustainable Transportation

Xue Wang and Rajesh Karki

Abstract This chapter titled "Reliability Assessment of Power Grid Supporting Sustainable Transportation" presents analytical models to evaluate the reliability of a power system with PHEV penetration considering the important PHEV characteristics and charging scenarios. The parameters of the PHEV-30 and the transportation data from NHTS 2009 were used to create the relevant models to obtain the load model of a PHEV. A MCS method was applied to combine individual PHEV models and build the overall model of PHEV fleet. The impact of PHEV penetration in a power system was assessed using the IEEE-RTS as a test system. Different PHEV penetrations and charging scenarios were examined to analyze the results. The results show that the system reliability degrades with the increase in PHEV penetration, and the negative impact is increased with fast charging which requires a higher power demand for a shorter duration. The negative reliability impacts can however be mitigated by shifting and controlling the charging time, which is area of ongoing research.

1 Introduction

Electric vehicle (EV) provides a means of transportation that utilizes sustainable energy, and provides an alternative to conventional transport vehicles that burn petroleum fuel and release harmful greenhouse gas emissions [1]. EV utilizes battery for electrical energy storage that can be generated from sustainable energy sources, such as wind and photovoltaic. EV can considerably reduce emissions that are believed to contribute to global warming [2]. There are different types of EV, such as hybrid electric vehicle (HEV), battery electric vehicle (BEV), and plug-in

X. Wang (✉) · R. Karki
Department of Electrical and Computer Engineering,
University of Saskatchewan, Saskatoon, Canada
e-mail: shaylacanada@gmail.com

R. Karki
e-mail: rajesh.karki@usask.ca

N.R. Karki et al. (eds.), *Sustainable Power Systems*, Reliable and Sustainable Electric Power and Energy Systems Management,
DOI 10.1007/978-981-10-2230-2_1

hybrid electric vehicle (PHEV). PHEV takes the concept of HEV a step further with a larger battery and gets rid of the perceived limitation of BEV in driving range, long charging time and dependence on electric outlets. Among all the other EV alternatives, PHEV receives increasing attention for lower emission and significantly less oil dependency. Moreover, the fuel-switching capability provides increased flexibility on driving range and economic electric propulsion [3].

Many countries have plans to substantially increase the number of PHEVs on the road by the next decade [4]. The advancement in PHEV technology has gradually established these vehicles as an affordable and efficient option capable of competing with traditional internal combustion engine vehicles. There is ongoing research in new types of energy storage technology, which provide increased compatibility with PHEV. Many researchers in academia and EV industry have focused on improving efficiency and reliability of the PHEV technology. An equally important area is to develop methodology to evaluate the efficiency and reliability of the overall energy system that generates and delivers energy to the electric transportation system. There is a need to develop appropriate models that recognize PHEV behavior, the power system to which the PHEVs are connected, and the interaction between the two PHEV and system models.

Many researchers have been involved in the study of various issues associated with PHEV injection in power systems since the first PHEV prototype was built. Existing literatures cover a broad range of PHEV technology research and applications. Reference [5] predicts the potential impacts of PHEVs on electricity demand and prices, generation structure, and associated emission levels. The work in [6] emphasizes on power quality and analyzes PHEV impacts on power losses as well as voltage deviations. PHEV impacts on a distribution system are analyzed in [7] using both deterministic and stochastic analytical methods considering PHEV penetration and the charging behaviors. Reference [8] presents PHEV impacts on a distribution system considering fast charging scenario. Reference [9] shows the improvement in power quality using coordinated charging realized by a smart metering system and also indicates the uncoordinated charging of PHEVs decreases the distribution system efficiency. Sustainable energy sources are considered in some literature related to PHEVs. Results of Ref. [10] show renewable energy sources can reduce the impacts of charging PHEVs on the distribution system in system losses as well as customers' cost. Reference [11] presents a system model with PHEVs and renewable energy resources, and analyzes related energy, economic and environmental impacts to power grid. The flexibility in the PHEV characteristics enables them to be charged when needed as long as electric outlet is available. When the number of PHEVs connected to a power system is considerable, the resulting change in the energy demand characteristic will adversely affect the overall system reliability. The reliability of a power system will be highly influenced by PHEV characteristics and operation strategies in the near future since PHEV penetration is expected to significantly increase. There is a definite need for further research in this area.

This chapter presents analytical models and a methodology to evaluate the reliability of a power system with PHEV penetration considering certain PHEV characteristics and charging scenarios. The commuting data provided by the United States transportation survey indicates that privately owned PHEVs account for a large proportion in current market and a majority of the people choose home to be the charging place at the end of the day. However, vehicles can also be charged in public parking places wherever the power outlet is available. In North America, PHEV models introduced by main automobile companies have two optional charging voltage levels: 120 and 240 V, which are also known as the normal charging and fast charging scenarios. The charging duration time at the higher voltage level is much shorter than the time required at the lower voltage level. Different scenarios of charging places and charging voltage levels are considered in this chapter.

This chapter is structured as follows. Important characteristics of PHEV modeling are presented in Sect. 2. Section 3 introduces the methodology of combining individual PHEV models and obtaining PHEV fleet model. Then charging scenarios of PHEVs are presented in Sect. 4. Section 5 introduces power system and load model that is used for PHEV reliability assessment, and followed by impacts of different PHEV penetration levels and charging scenarios on power system reliability.

2 Important PHEV Characteristics for Model Development

PHEV is a hybrid electric vehicle in which a gasoline engine and an electric motor are both used for propulsion. The vehicle includes rechargeable battery packs that can be charged as needed through an external electric power source. PHEV is usually operated at all-electric mode until the battery reaches the minimum state of charge (SOC). Then gasoline becomes the main energy source to drive the vehicle, and battery SOC may fluctuate due to regenerative braking. Lithium-ion batteries are widely used in PHEVs and are normally sized to provide enough capacity for short commutes.

PHEV-X refers to PHEVs with an all-electric range of X miles. Table 1 provides the important parameters of PHEV-30, which is a typical model used by many manufactures. Similar models are also used by several research institutes, such as the Sloan Automotive Laboratory at MIT (U.S.), United States Advanced Battery

Table 1 Parameters of PHEV-30

Body size	Midsize vehicle
Battery type	Lithium-ion
Battery size (kWh)	8
Full recharge time (120 V outlet)	7 h
Full recharge time (240 V outlet)	2.5–3 h
All-electric range (Miles)	30

Consortium, and National Renewable Energy Laboratory [12–14]. The following subsections describe the PHEV parameters that are important for reliability and energy modeling.

2.1 Daily Driving Distance

PHEV behavior is determined in general by several factors, such as, transportation habits, charging infrastructure, price incentives, and the battery technologies [4]. The daily driving distance of a vehicle is an important parameter that can vary considerably from one day to another. Daily vehicle travel data have been collected and reported by the Federal Highway Administration of U.S. Department of Transportation. The latest publication with this data was collected in the National Household Travel Survey (NHTS) 2009 [15, 16]. Data from the report can be used to model the uncertainty in the driving distance. An analysis of the data shows that the daily driving distance can be represented by a lognormal distribution using a best-fit criterion. This is shown in Fig. 1 and the lognormal distribution is mathematically expressed in Eq. (1).

$$f_D(x) = \frac{1}{x\sigma_D\sqrt{2\pi}} e^{\left[-(\ln x-\mu_D)^2/2\sigma_D^2\right]} \tag{1}$$

where
μ_D is the mean of the log-normal of the daily driving distance, equal to 4.0, and
σ_D is the standard deviation, equal to 0.92.

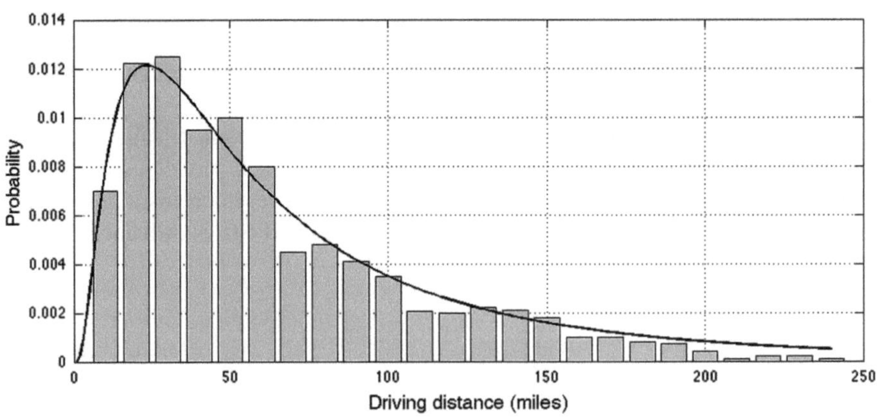

Fig. 1 Probability of daily driving distance

2.2 Battery Performance During Driving

A fully charged PHEV will initially operate in the "all-electric mode" using the electric power provided by the battery as described in Sect. 1. The SOC is another parameter of a PHEV that indicates the amount of electric energy available in the battery at any instant in time. As the vehicle is operated, the battery is gradually discharged, and the value of SOC declines until it reaches the minimum value about 0.2, which means 20 % of the full capacity. When the minimum battery SOC is reached, the vehicle is switched to the "gasoline mode." A toolbox called ADVISOR operated in MATLAB/Simulink environment has been used to assess the battery performance during the driving period of a day. Figure 2 shows the battery SOC as a function of the driving distance.

The plot of SOC with driving distance in Fig. 2 can be divided into two distinct periods; the initial "all-electric" driving period and the "gasoline" driving period. The initial period is battery dependent, and the SOC declines linearly with driving distance. In the case of PHEV-30, the first stage will approximately last for 30 miles. Equation (2) gives the battery SOC at this stage.

$$\mathrm{SOC} = \begin{cases} \max\{0.2, 1 - 0.0333D\}, D \leq X \\ Y, D > X \end{cases} \tag{2}$$

where D is the daily driving distance in miles, X is the all-electric driving range in miles, and Y is a uniformly distributed random number between 0.2 and 0.3.

The second stage of the driving period is associated with the "gasoline mode." During this stage, the PHEV battery SOC fluctuates as the battery is frequently charged for short durations due to regenerative braking, which allows restoring energy whenever a PHEV slows down. The battery is discharged to provide energy

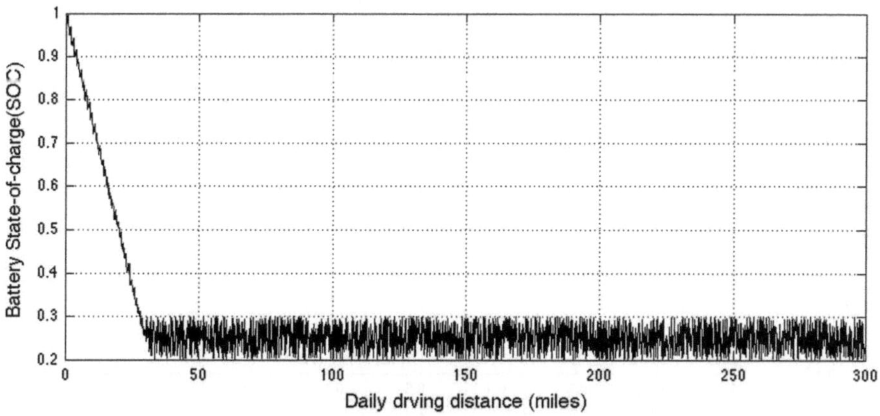

Fig. 2 Battery SOC as a function of driving distance

as the vehicle speeds up again. The battery SOC varies in a stochastic manner in this stage. A uniformly distributed random number between 0.2 and 0.3 can be used to obtain the SOC value in the second stage of the driving period. The battery performance model during the vehicle driving period can therefore be achieved by combining the respective models for the two stages.

2.3 Battery Performance During Charging

Most people start charging their vehicles as soon as they arrive home at the end of a day [16]. The PHEVs would then be plugged in for battery charging if the battery SOC is below a specified value, such as 0.8. The battery is charged until it is full. The battery SOC increases almost linearly with charging time. Figure 3 shows the battery SOC as a function of charging time for a typical lithium-ion battery. Figure 4 shows power supplied to the battery during the charging process in per unit of the rated charging power.

A PHEV plugged in for charging can be considered as a power system load with a specific characteristic. The load profile of a PHEV is characterized by the charging power and duration. The duration on the PHEV load is the time taken to fully charge the battery, and can be calculated using Fig. 3 with the knowledge of the initial battery SOC prior to the charging process. The power demand of the PHEV load as a function of the charging time can be calculated using Fig. 4.

The Charging Start Time for PHEV battery charging is an important parameter in PHEV load modeling. Daily travel data collected and reported in NHTS 2009 can be used to obtain the "last trip ending time" for a PHEV for each day. Since it is assumed that the PHEV charging will commence as soon as the vehicle reaches home, the last trip ending time is also the charging start time. The probability distribution of the "last trip ending time" data is shown in Fig. 5. The figure also

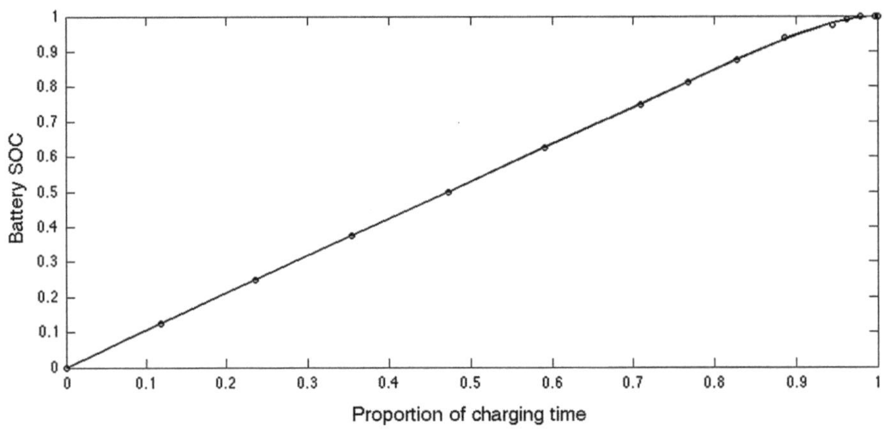

Fig. 3 Battery SOC during the charging process

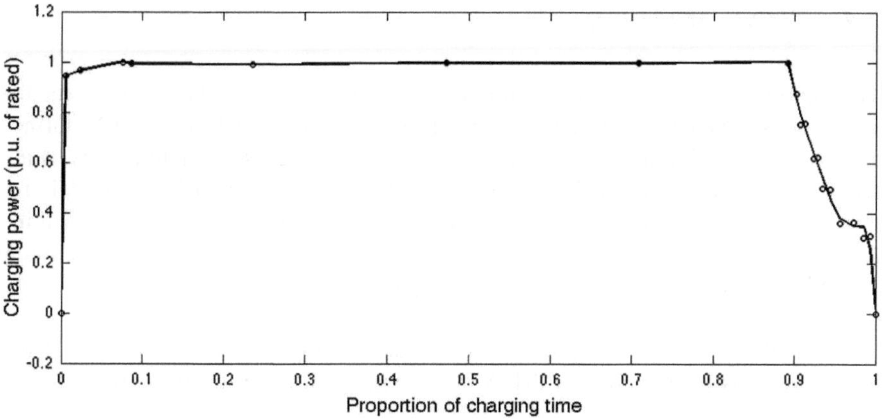

Fig. 4 Power demand of a PHEV during the charging process

shows the best fitting curve which is a normal distribution. The mathematical equation for this curve is given in Eq. (3).

$$f_{T_h}(x) = \begin{cases} \frac{1}{\sigma_{T_h}\sqrt{2\pi}} e^{\left[-(x-\mu_{T_h})^2/2\sigma_{T_h}^2\right]}, (\mu_{T_h} - 12) < x \leq 24 \\ \frac{1}{\sigma_{T_h}\sqrt{2\pi}} e^{\left[-(x+24-\mu_{T_h})^2/2\sigma_{T_h}^2\right]}, 0 < x \leq (\mu_{T_h} - 12) \end{cases} \quad (3)$$

where

μ_{Th} Mean value, equals to 17.4 h, or 5:24 PM

σ_{Th} Standard deviation, equals to 3.3 h

T_h start time of the home charging scenario

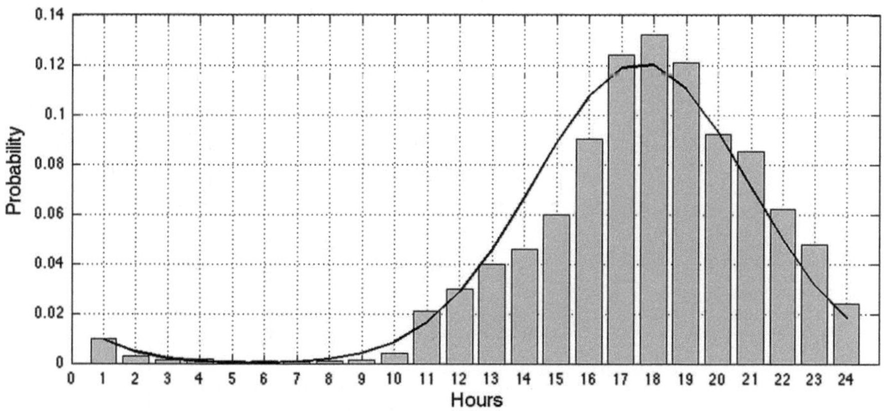

Fig. 5 Probability distribution of charging start time for home charging

2.4 PHEV Modeling

The important PHEV parameters and their characteristics described in the preceding subsections are analyzed to develop an appropriate load model for a PHEV during a charging cycle. The developed load model can then be used for reliability evaluation. Figure 6 presents an algorithm that can be used to obtain the load model of an individual PHEV for one charging cycle. The number of charging cycles within a day depends on the frequency of the PHEV charging processes available in a day. There is only one charging cycle in a day if charging at home is the only option that occurs at the end of the day.

Each block in Fig. 6 represents a model development for a specific PHEV parameter. It is assumed that a PHEV will be plugged in for charging if the battery

Fig. 6 Load model development for an individual PHEV

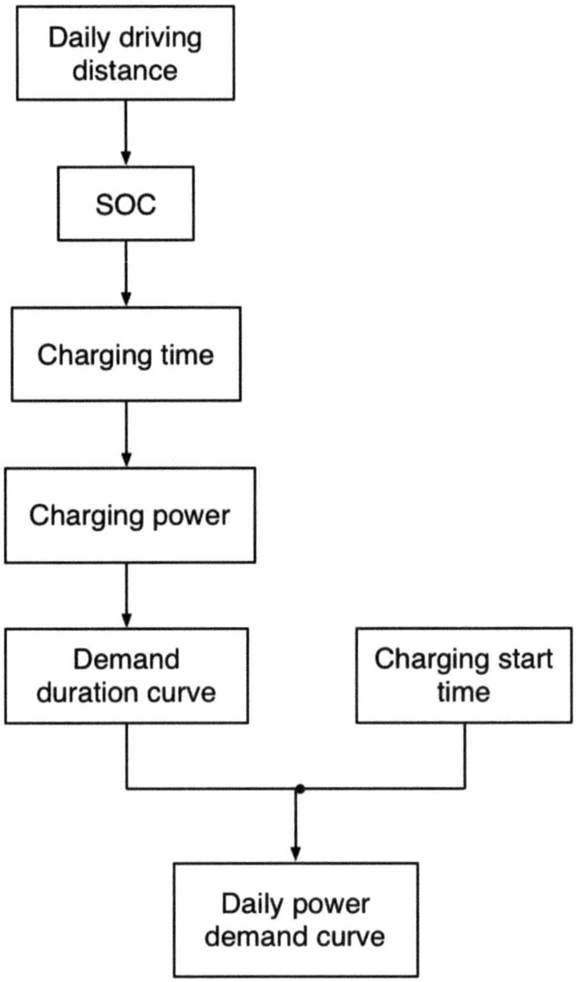

SOC is lower than 0.8. The battery SOC can be calculated from the knowledge of the daily driving distance using the relation in Fig. 2 and Eq. (2). If the calculated SOC is less than 0.8, the PHEV will be plugged in for battery charging. It is important to record the charging start time. The charging duration or the time taken to fully charge the PHEV battery is then obtained. The amount of power supplied to the battery for the charging duration is then determined, which then provides the load or demand profile of an individual PHEV in time chronology. If there is more than one charging cycle within a day for the PHEV, the load profiles are similarly created for each charging cycle, and a time sequential load profile is obtained for the entire day.

3 PHEV Fleet Modeling

Section 2 presents the development of the load model for an individual PHEV considering a daily charging cycle. PHEV fleet consists of a number of PHEVs with different charging and driving behaviors. These parameters are stochastic and interrelated with each other. The model developed for individual PHEVs within a fleet need to be aggregated using an appropriate technique to obtain the characteristic model for the overall fleet.

A Monte Carlo Simulation (MCS) method is used to build the overall model for a PHEV fleet by combining individual PHEV models. Figure 6 shows how the important parameters of an individual PHEV are sequentially modeled to obtain the daily load profile of the PHEV. The sequential modeling is repeated for all the PHEVs that exist in the fleet using the MCS technique to obtain the overall daily load profile for the fleet.

In the beginning of the simulation, a random number Z_i is generated to obtain the daily driving distance D_i of the ith PHEV. The random number Z_i is used to calculate a lognormal value M_i of the driving distance, using Eq. (4) where the parameters μ_D and σ_D are given in Eq. (1).

$$M_i = \mu_D + \sigma_D * Z_i \tag{4}$$

The calculated value of M_i is then used in Eq. (5) to obtain the daily driving distance D_i of the ith PHEV.

$$D_i = e^{M_i} \tag{5}$$

The value of D_i obtained from Eq. (5) is then used in Eq. (2) to calculate the SOC_i of the ith PHEV. The SOC value is then fitted in Fig. 3 to determine the time required to fully charge, or the charging time CT_i of the ith PHEV.

In order to simulate the charging start time T_i for the ith PHEV, a uniformly distributed random number U_i between 0 and 1 is first generated. The random number U_i is transformed to a normally distributed variable Y_i using Eqs. (6) and (7).

The charging start time T_i is then calculated using Eq. (8) where μ_{Th} and σ_{Th} are obtained from Eq. (3).

$$Q_i = \begin{cases} 1 - U_i, 0.5 < U_i \leq 1.0 \\ U_i, 0 \leq U_i \leq 0.5 \end{cases} \tag{6}$$

$$Y_i = \begin{cases} a_i, 0.5 < U_i < 1.0 \\ 0, U_i = 0.5 \\ -a_i, 0 \leq U_i < 0.5 \end{cases} \tag{7}$$

where

$$b_i = \sqrt{-2\ln Q_i}$$

and

$$a_i = b_i - \frac{2.515517 + 0.802853 b_i + 0.010328 b_i^2}{1 + 1.432788 b_i + 0.189269 b_i^2 + 0.001308 b_i^3}$$

$$T_i = \mu_{Th} + \sigma_{Th} Y_i \tag{8}$$

The calculated SOC_i and charging time CT_i is used in Fig. 4 to obtain a time sequential power demand of the ith PHEV for the charging time duration. This creates the sequential load profile for which the start time is T_i and duration is CT_i. This provides the load model for the ith PHEV for a charging cycle.

To effectively specify this sequential modeling process, one illustrational charging cycle is shown in Fig. 7.

The sequential modeling process is carried out for all the n PHEVs in the fleet as shown in Fig. 8.

The daily sequential loads obtained as discussed above are recursively aggregated for all the n vehicles as shown in Eq. (9) to obtain the sequential load model for the entire fleet.

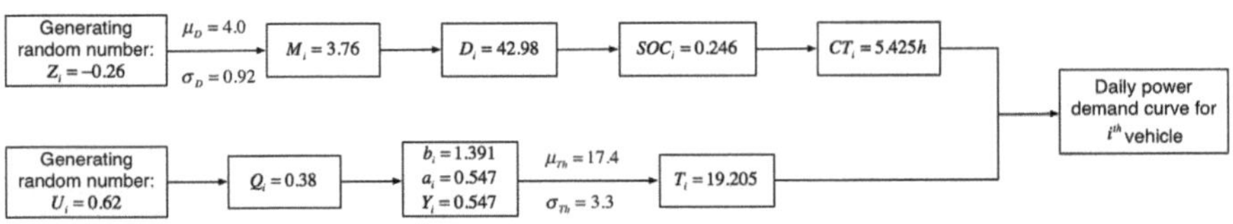

Fig. 7 A sample run of the PHEV charging cycle methodology

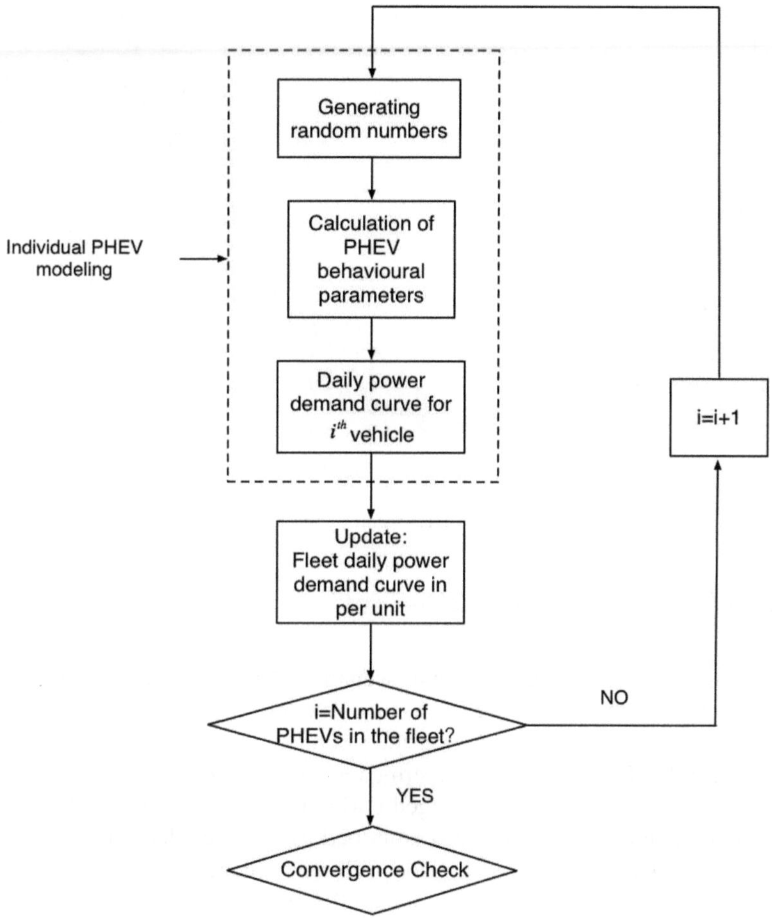

Fig. 8 PHEV fleet modeling procedure

$$L_h = \sum_{i=1}^{n} L_{\mathrm{PHEV}_{i,h}} \tag{9}$$

where, $h \in 24$ is the hour of the day, L_h is the fleet load at Hour h, and $L_{\mathrm{PHEV}_{i,h}}$ is the load of the ith PHEV at Hour h.

The simulation runs for all the vehicles in fleet are repeated until convergence criteria are met, i.e., when the fleet load model is not changed with further runs. Figure 9 shows the daily sequential load model for a fleet with 100 PHEVs and the individual updated load models in per unit of the respective peak loads.

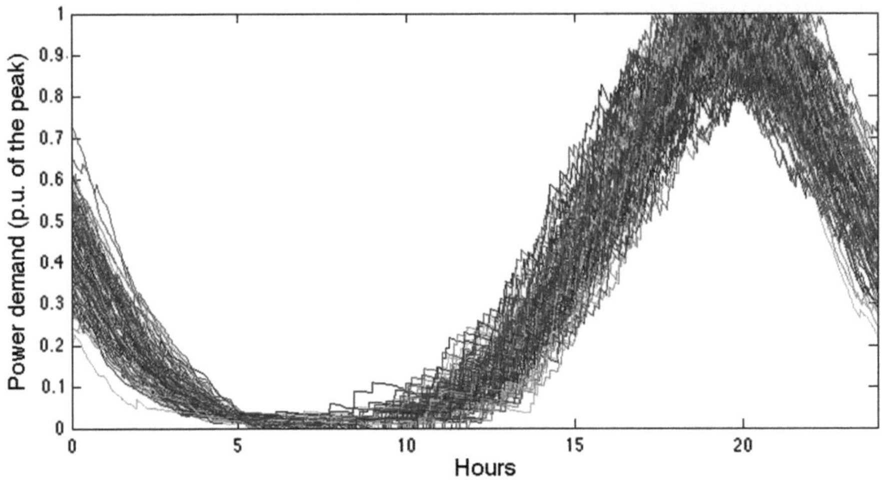

Fig. 9 Daily load model of individual PHEVs and of the entire fleet

4 PHEV Charging Scenarios

There can be different methods and scenarios for PHEV charging. Section 3 illustrated the method to develop a load model for a PHEV fleet with the assumption that people start charging their vehicles as soon as they arrive home. This scenario for PHEV charging is termed as "Home Charging Only" scenario in this chapter. PHEVs can also be charged during the daytime where public charging facilities are available. This scenario is termed as "Public Charging."

The PHEV battery can be recharged at different voltage levels as specified by the PHEV manufacturer. Table 1 shows that the PHEV-30 can be charged either at 120 and 240 V, which are designated as "normal charging" and "fast charging" methods. The charging time depends on the charging voltage level, and therefore the charging voltage will influence the load model of the PHEV. The load model illustrated in Fig. 9 was obtained considering a normal charging voltage of 120 V.

4.1 Normal and Fast Charging

Table 1 shows that the normal charging of PHEV-30 using 120 V supply requires 7 h for full recharge. The table also shows that the full recharge time is only 2.5–3 h when fast charging mode is selected using 240 V supply. Among PHEV models introduced by main automobile companies of North America, fast charging is a common feature that provides the additional flexibility in the charging process. It should be noted the charging time is reduced and the charging power is increased when fast charging method is selected to charge the PHEV. In this case, the relevant

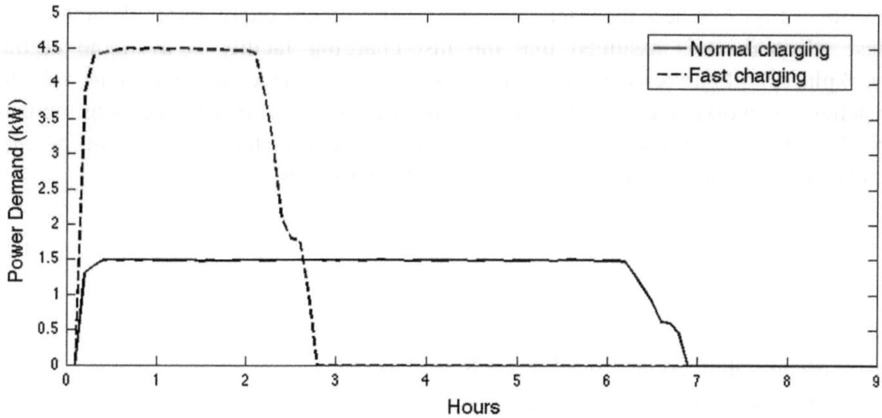

Fig. 10 Comparisons of the power demand and charging time of the normal and the fast charging methods

full recharge time and charging power should be used in Figs. 3 and 4 to obtain the PHEV load model.

Figure 10 compares the load model of a PHEV using the two different charging methods, i.e., the normal charging and the fast charging. It can be seen that the fast charging method requires less than half the charging time and about three times higher charging power when compared to the normal charging method.

4.2 Home and Public Charging

It can be assumed that the majority of vehicles that have relatively short daily driving distances will prefer the "home charging only" scenario. This in case, the normal charging method using 120 V supply is the preferred option as this voltage level is readily available in homes.

PHEV owners can also charge their vehicles during the daytime where charging facilities are available. It is reasonable to assume that people would choose fast charging in public during daytime for the sake of time saving. It is highly likely that PHEV subjected to public charging will also be recharged at home at the end of the day. These vehicles will have at least two charging cycles in a day. When public charging scenario is considered, individual PHEV load model can be obtained by evaluating the charging cycle shown in Fig. 6 twice considering two charging cycles within a day.

The following assumptions are made to model the *i*th PHEV under public charging. First, the driving distance of a PHEV prior to public charging is assumed to be half of daily driving distance, which is 0.5 * D_i. The charging time is then calculated using the relationship in Fig. 3 under fast charging scenario considering 3 h as full recharge time. The charging power is calculated using the per unit power profile in Fig. 4, where the rated charging power for fast charging is selected as the base. Second, it is assumed that the fast charging facility is available at the workplace, and public charging of the PHEV will commence as soon as the vehicle reaches the workplace. The travel data of "arriving workplace time" reported in NHTS 2009 can be used to obtain the "public charging start time." Using best-fit criterion, the mathematically expression is shown in Eq. (10).

$$f_{Tp}(x) = \frac{1}{\sigma_{Tp}\sqrt{2\pi}} e^{\left[-(x-\mu_{Tp})^2/2\sigma_{Tp}^2\right]} \tag{10}$$

where

μ_{Tp} Mean value, equals to 8.5 h, or 8:30 AM

σ_{Tp} Standard deviation, equals to 2.4 h

Tp start time of public charging scenario.

The fully charged PHEV at the workplace will then be discharged on the way home. The driving distance prior to home charging is assumed to be half of daily driving distance, which is $0.5 * D_i$. Home charging cycle evaluation for the PHEV is carried out using the method discussed in Sect. 3 considering the normal charging method. Two load models are therefore obtained for the two charging cycles of each PHEV, i.e., public charging and home charging cycles. Then as described in Sect. 3, the PHEV fleet model can be obtained using MCS method.

Figure 11 shows the daily sequential load model for a fleet with 100 PHEVs under "public/home charging" scenario and the individual updated load models in per unit of the respective peak loads. The daily load model shown in Fig. 11 can be compared with Fig. 9 obtained for the "home charging only" scenario. It can be observed from Fig. 11 that there is a new peak during daytime introduced due to public charging, and the peak value during the daytime is higher than the evening "home charging" peak because of the higher charging power needed for public charging as shown in Fig. 10.

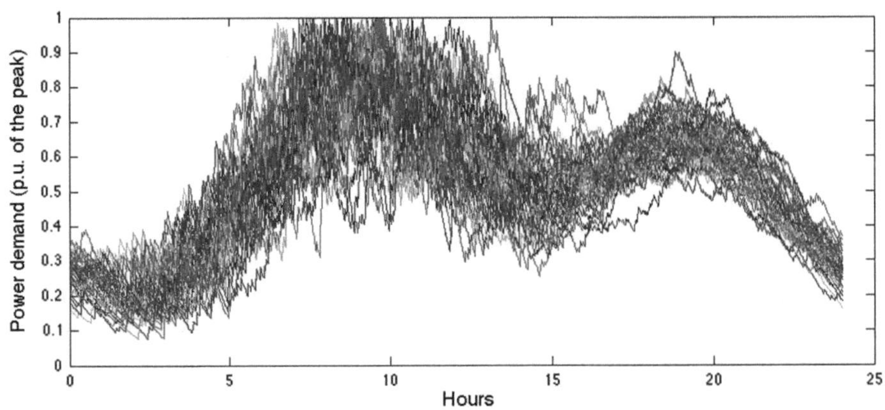

Fig. 11 Daily load model of individual PHEVs and of the entire fleet for the public/home charging scenario

5 Power System Reliability Modeling to Incorporate PHEV Loads

The IEEE Reliability Test System (IEEE-RTS) is used in this chapter to illustrate the reliability impacts of PHEV penetration in electric power systems. The IEEE-RTS is a relatively large power system model with a total installed capacity of 3405 MW and a peak load of 2850 MW. The annual hourly chronological load for the IEEE-RTS expressed in per unit of the peak load is shown in Fig. 12. The annual peak load occurs during the winter season. Power demand is greater during wintertime because of the electrical heating load, while not has much impact during summer.

A period analysis is carried out by dividing a year into two seasons, in order to incorporate the seasonality effect on the system reliability. Summer runs from April to September and contains 4392 h. Winter consists of 4368 h, from October to March. Figure 13 shows the 24-h chronological load for the winter day with the highest annual load and the summer day with the lowest load. The two days are respectively from December and September as shown in Fig. 12.

The original IEEE-RTS described above is modified by adding PHEV load to the test system. The magnitude of the PHEV load relative to the original system load is expressed in terms of PHEV penetration. It is the ratio of the maximum PHEV power demand and the system peak load. For example, a 10 % PHEV penetration in the IEEE-RTS means a maximum PHEV demand of 285 MW. The total system load of the modified IEEE-RTS is obtained by aggregating the original load with the PHEV load.

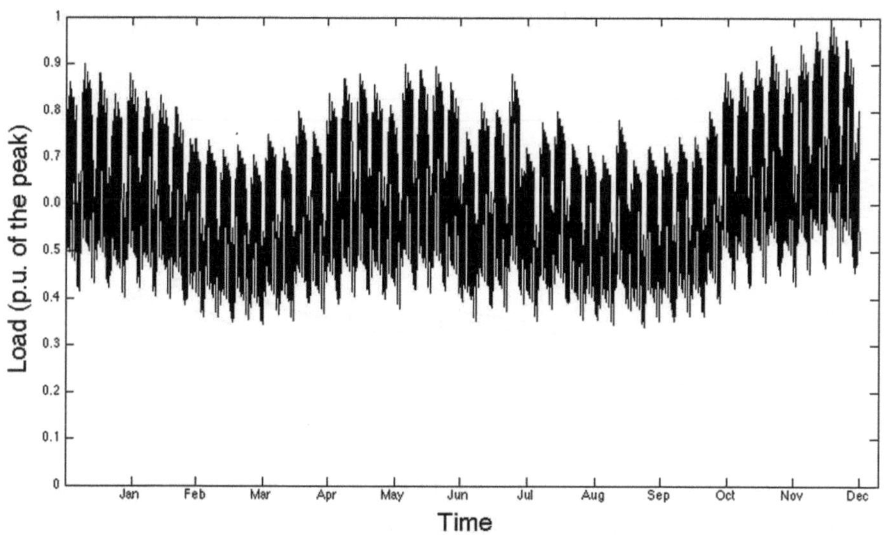

Fig. 12 The IEEE-RTS chronological hourly loads for 1 year

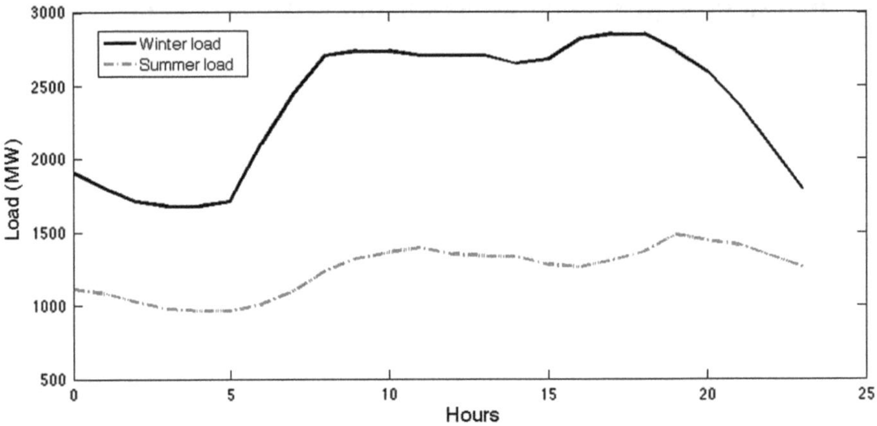

Fig. 13 The IEEE-RTS daily load curves with the highest and lowest annual loads

Figure 14 shows the summer daily load models obtained by aggregating the original load with the PHEV load considering 25 % PHEV penetration. Figure 14 shows the load models for the two charging scenarios: "home charging only," and "public/home charging." In the second scenario, it is assumed that all the PHEV can be charged at home and only 40 % of the PHEV can also be charged through public charging. The original load is also shown for comparison. The horizontal line indicates the annual peak load of the original IEEE-RTS. Similarly, Fig. 15 shows the original and modified daily load models for the winter day for the two PHEV charging scenarios.

It can be observed from Figs. 14 and 15 that the overall system power demand is significantly increased due to PHEV charging during the peak hours (approximately

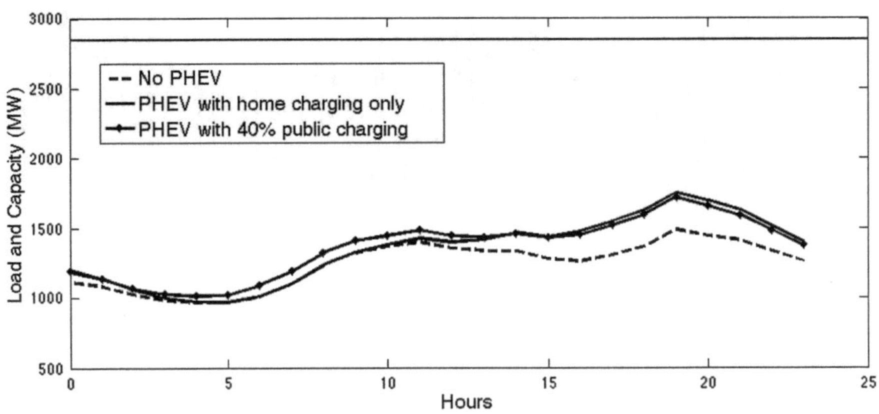

Fig. 14 Summer daily load models with and without PHEV

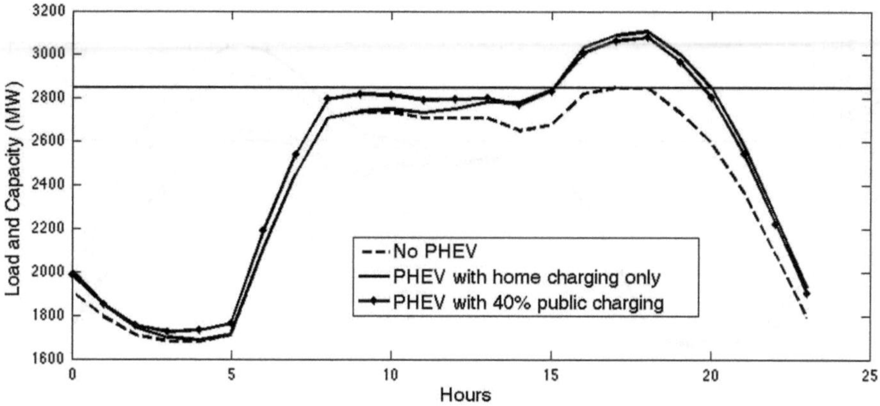

Fig. 15 Winter daily load models with and without PHEV

from 16:00 to 21:00) of the day. The peak load value substantially exceeds the original annual peak load of 2850 MW during the winter season, which indicates the overlap of PHEV fleet charging demand and system original household demand.

A comparison of the two charging scenarios in Figs. 14 and 15 shows that the "home charging only" scenario has a greater impact on increasing the annual peak load. The annual peak load is also increased significantly in the "public and home charging" scenario. This scenario, however, utilizes additional power throughout the day including the "valley" period in the morning.

Figure 16 shows the impact of PHEV penetration on the system daily load model during the summer season considering the home charging only scenario. The PHEV penetration is increased from 5 to 50 %. Figure 17 shows the impact on the winter load model. The two horizontal lines indicate the annual peak load of the original IEEE-RTS, and the dashed horizontal line indicates the total installed capacity of the system.

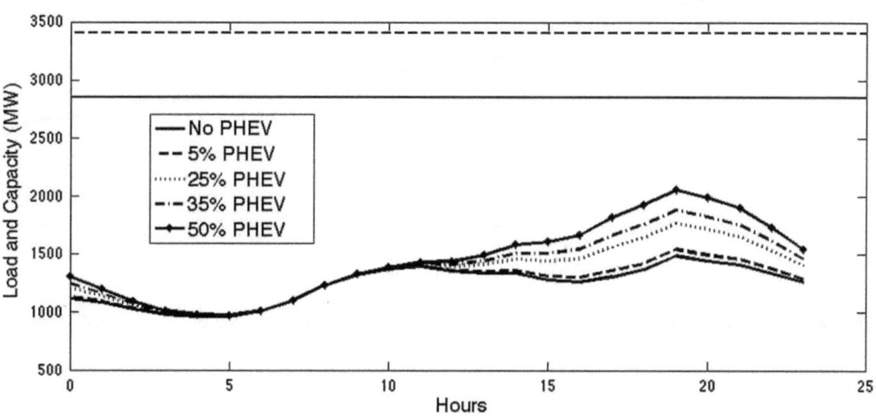

Fig. 16 The impact of PHEV penetration on the summer daily load model

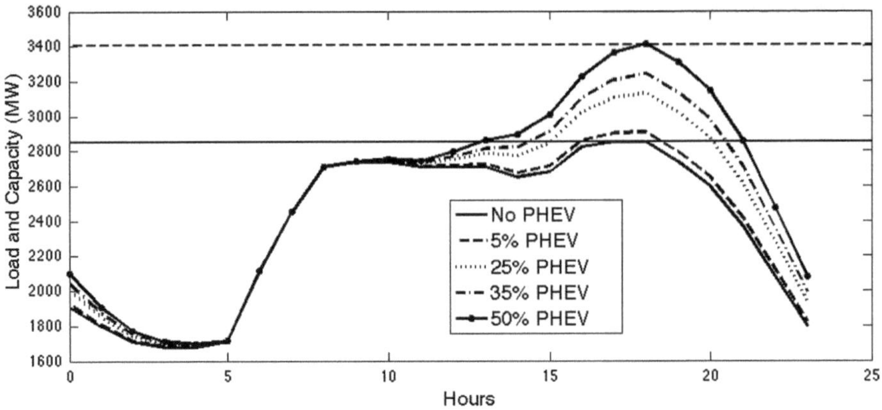

Fig. 17 The impact of PHEV penetration on the winter daily load model

The method of aggregating the 24-h PHEV load model with the system chronological load is repeated for each in the summer and winter seasons, and the aggregated hourly loads for each season are sorted in a decreasing order to obtain the modified load duration curve (LDC) for each season. Figure 18 shows the winter LDC of the IEEE-RTS with different PHEV penetration levels considering the "home charging only" scenario. It can be seen that there is a 20 % increase in the annual peak load with 50 % PHEV penetration.

The generation model for the system is created for each seasonal period in the form of capacity outage probability tables [17]. The generation model is convolved

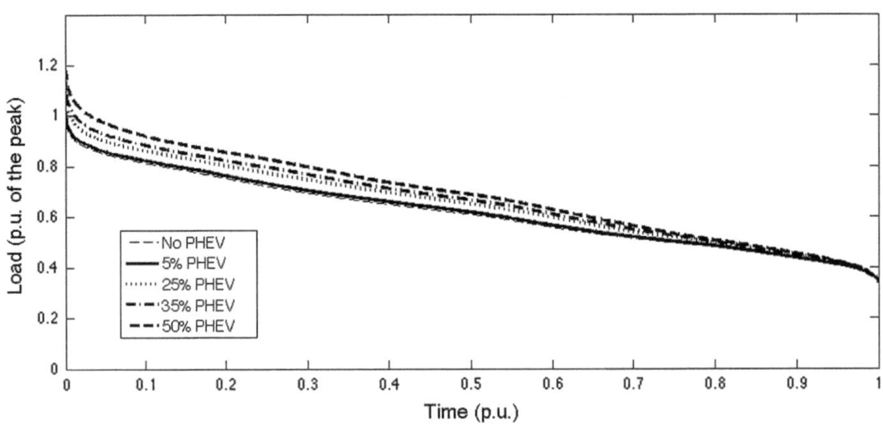

Fig. 18 Winter LDC with varying PHEV penetration considering "home charging only" scenario

with the load models for the respective season above to obtain the risk indices for each season. The seasonal indices are combined to obtain the annual indices, such as the loss of load expectation (LOLE) and loss of energy expectation (LOEE).

6 Impact of PHEV on Power System Reliability

This section analyzes the reliability impacts of PHEV penetration on a power system. A reliability study of the IEEE-RTS was carried out by considering different levels of PHEV penetration into the system. The reliability indices were evaluated while varying the PHEV penetration from 5 to 50 %. The "home charging only" scenario was considered in this study. Table 2 shows that the system LOLE and LOEE increase considerably as the PHEV penetration increases.

A study was also carried out to assess the impacts of different PHEV charging scenarios on the reliability indices. The reliability indices of the IEEE-RTS with 25 % PHEV penetration were evaluated for the two PHEV charging scenarios: "home charging only" and "home/public charging." It was assumed that all the PHEV were charged at home and only 40 % of them could be charged in public in the "home/public charging" scenario. Table 3 shows the LOLE and the LOEE for the two scenarios. It can be seen that both the indices are higher in the "public/home charging" than in the "home charging only" scenario.

Table 2 System reliability indices with increasing PHEV penetration considering home charging only

PHEV penetration levels (%)	LOLE (h/year)	LOEE (MWh/year)
0	9.4180	1.174
5	11.1913	1.425
25	25.4625	3.593
35	40.0766	5.949
50	76.8578	12.694

Table 3 System reliability indices for different charging scenarios

PHEV charging scenario	LOLE (h/year)	LOEE (MWh/year)
No PHEV	9.4180	1.174
25 % PHEV Home charging only	25.4625	3.593
25 % PHEV Public/home charging	26.9852	3.749

7 Conclusions

This chapter presents analytical models to evaluate the reliability of a power system with PHEV penetration considering the important PHEV characteristics and charging scenarios. The parameters of the PHEV-30 and the transportation data from NHTS 2009 were used to create the relevant models to obtain the load model of a PHEV. A MCS method was applied to combine individual PHEV models and build the overall model of PHEV fleet. The impact of PHEV penetration in a power system was assessed by using the IEEE-RTS as a test system. Different PHEV penetrations and charging scenarios were examined to analyze the results. The results show that the system reliability degrades with PHEV penetration increases, and the negative impact is increased with fast charging which requires a higher power demand for a shorter duration. The negative reliability impacts can however be mitigated by shifting and controlling the charging time, which is area of ongoing research.

References

1. Knipping E, Duvall M (2007) Environment assessment of plug-in hybrid electric vehicles volume 1: nationwide greenhouse gas emissions. Electric Power Research Institute and Natural Resources Defense Council
2. Elgowainy A, Burnham A, Wang M, Molburg J, Rousseau A (2009) Well-to-wheels energy use and greenhouse gas emissions analysis of plug-in hybrid electric vehicles. Energy System Division, Argonne National Laboratory
3. Li Y (2007) Scenario-based analysis on the impacts of plug-in hybrid electric vehicles' (PHEV) penetration into the transportation sector. In: IEEE International symposium technology and society, ISTAS 2007
4. Lojowska A, Kurowicka D, Papaefthymiou G, van der Sluis L (2011) From transportation patterns to power demand: stochastic modeling of uncontrolled domestic charging of electric vehicles. In: IEEE power and energy society general meeting
5. Hadley SW, Tsvetkova A (2008) Potential impacts of plug-in hybrid electric vehicles on regional power generation. Oak Ridge National Laboratory, U.S. Department of Energy
6. Green II RC, Wang L, Alam M (2011) The impacts of plug-in hybrid electric vehicles on distribution networks: a review and outlook. Renew Sustain Energy Rev 15:544–553
7. Taylor J, Maitra A, Alexander M, Brooks D, Duvall M (2009). Evaluation of the impact of plug-in electric vehicle loading on distribution system operations. In: IEEE power and energy society general meeting
8. Yunus K, De La Parra HZ, Reza M (2011). Distribution grid impacts of plug-in electric vehicles charging at fast charging stations using stochastic charging model. In: Proceedings of the 2011–14th European conference power electronics and applications
9. Clement-Nyns K, Haesen E, Driesen J (2010) The impact of charging plug-in hybrid electric vehicles on a residential distribution grid. IEEE Trans Power Syst 25(1):371–380
10. Hafez O, Bhattacharya K (2012) Optimal PHEV charging in coordination with distributed generation operation in distribution systems. In: IEEE power and energy society general meeting
11. Marano V, Rizzoni G (2008) Energy and economic evaluation of PHEVs and their interaction with renewable energy sources and the power grid. In: Proceedings of IEEE international conference on vehicular electronics and safety

12. Kromer MA, Heywood JB (2007) Electric powertrains: opportunities and challenges in the U. S. light-duty vehicle fleet. Sloan Automotive Laboratory. Laboratory for Energy and the Environment Massachusetts Institute of Technology
13. Axsen J, Burke A, Kurani K (2008) Batteries for plug-in hybrid electric vehicles (PHEVs): goals and the state of technology circa 2008. Institute of Transportation Studies, University of California, California
14. Pesaran A (2007) Battery requirements for plug-in hybrid electric vehicles-analysis and rationale. National Renewable Energy Laboratory
15. NHTS Data Center, 2009 NHTS- Version 2.1: http://nhts.ornl.gov/download.shtml#2009
16. U.S. Department of Transportation, Federal Highway Administration (2009) Summary of travel trends-2009 National Household Travel Survey
17. Billinton R, Allan RN (1996) Reliability evaluation of power systems. Plenum Press, New York

Chapter 2
Hybrid AC/DC Micro-Grids: Solution for High Efficient Future Power Systems

Peng Wang, Jianfang Xiao, Chi Jin, Xiaoqing Han and Wenping Qin

Abstract This chapter titled "Hybrid AC/DC Microgrids: Solution for High Efficient Future Power Systems" presents a new configuration for future power systems which is the hybrid AC/DC gird for high efficient connection of the inherent AC and DC sources and loads. Three-phase AC power systems have been in dominant position for over hundred years due to invention of transformer and the inherent characteristic from fossil energy-driven rotating machines. However, the gradual changes of load types and distributed renewable generation (DRG) in AC local distribution systems provide food for consideration of adding DC networks. Renewable sources such as fuel cells and solar photovoltaics are DC inherent and should be connected to AC grid through DC/AC conversion techniques whereas some AC inherent renewable sources like wind generators also need DC links in their conversion systems to increase efficiency and mitigate power variation caused by intermittency and uncertainty. The disadvantage of AC grids for connection of DC inherent sources and loads as well as AC loads with DC links is that additional DC/AC or AC/DC converters are required, which may result in efficiency loss from the reverse conversion. In the other hand DC grids are resurging due to the development and deployment of renewable DC power sources and their inherent advantage for DC loads in commercial, industrial and residential applications. The number of power conversions in a DC microgrid has been significantly reduced to enhance system energy efficiency. A more likely scenario is the coexistence of both

P. Wang (✉) · J. Xiao · C. Jin
Nanyang Technology University, Singapore, Singapore
e-mail: epwang@ntu.edu.sg

J. Xiao
e-mail: XI0012NG@e.ntu.edu.sg

C. Jin
e-mail: Jinchi@e.ntu.edu.sg

X. Han · W. Qin
Taiyuan University of Technology, Shanxi, China
e-mail: hanxiaoqing@tyut.edu.cn

W. Qin
e-mail: qinwenping1027@163.com

N.R. Karki et al. (eds.), *Sustainable Power Systems*, Reliable and Sustainable Electric Power and Energy Systems Management,
DOI 10.1007/978-981-10-2230-2_2

AC and DC microgrids, which is so-called the hybrid AC/DC microgrid in order to reduce processes of multiple reverse conversions in an individual AC or DC microgrid and facilitate the connection of various renewable AC/DC sources and loads to power system. Therefore the concept of hybrid microgrids, which can harmonize both AC and DC sources and loads, has been proposed for future high efficient power systems. Conventional AC and DC grids are interconnected together through the bidirectional AC/DC converter. The component model has been introduced. The control and operation of individual sources and energy storages are presented. The coordination control and power sharing techniques are also introduced.

1 Introduction

Three-phase AC power systems have been in dominant position for over 100 years due to invention of transformer and the inherent characteristic from fossil energy driven rotating machines. However, the gradual changes of load types and distributed renewable generation (DRG) in AC local distribution systems provide food for consideration of adding DC networks [1].

The limited fossil energy sources are reducing significantly due to the fast growth of energy demand in modern societies. Over dependent on fossil courses has caused severe environment problems such as air pollution, globe warming and extreme weather conditions. Therefore renewables like wind and solar as the alternatives of fossil sources have been integrated into power systems in last decades. Renewable sources can be connected to either the high voltage network of an AC grid such as wind solar farms or to the low voltage distribution networks as distributed generations like small backyard wind turbines and rooftop photovoltaic (PV) arrays. This is a major reason behind the appearance of AC microgrids [2–5]. However some renewable sources such as fuel cells and solar photovoltaics are DC inherent and should be connected to AC grid through DC/AC conversion techniques. Some AC inherent renewable sources like wind generators also need DC links in their conversion systems to increase efficiency and mitigate power variation caused by intermittency and uncertainty.

In the early stage, power systems were designed to supply the lighting, heating, and motor driving loads which are mainly AC type. However, load evolution in AC local distribution systems have been occurring quietly with the development of power electronics techniques and new lighting equipment for high efficiency of energy utilization and control flexibility. When we look around the loads in modern power systems, it is found that DC loads and AC loads with AC/DC/AC converters are in a dominate position. Common facilities such as computers, printers, videos recorders, TVs, microwave in offices and homes are direct DC loads. Traditional AC motors as direct drivers for washing machines, refrigerators, air conditioners and various industrial machines are being gradually replaced by power electronics based AC motors in order to control the motor speed and to save energy.

The disadvantage of AC grids for connection of DC inherent sources and loads as well as AC loads with DC links is that additional DC/AC or AC/DC converters are required, which may result in efficiency loss from the reverse conversion. Recently, DC grids are resurging due to the development and deployment of renewable DC power sources and their inherent advantage for DC loads in commercial, industrial, and residential applications. The DC microgrid has been proposed to integrate various distributed generators [6–11]. The number of power conversions in a DC microgrid has been significantly reduced to enhance system energy efficiency [1]. DC microgrid has been implemented for telecommunication system, data center, offshore platforms, renewable energy system, etc. However, AC sources have to be converted into DC before connected to a DC grid and DC/AC inverters are required for conventional AC loads. A more likely scenario is the coexistence of both AC and DC microgrids, which is so-called the hybrid AC/DC microgrid in order to reduce processes of multiple reverse conversions in an individual AC or DC microgrid and facilitate the connection of various renewable AC/DC sources and loads to power system. Therefore the concept of hybrid microgrids, which can harmonize both AC and DC sources and loads, has been proposed [1, 3] for future high efficient power systems. This chapter introduces basic network configuration. The major components and their functions will be presented. The control and operation of the major components will be discussed in detail to present the operating principle of the hybrid microgrid.

2 Hybrid AC/DC Microgrid Configuration and Components

A hybrid AC/DC microgrid is the combination of AC and DC networks as shown in Fig. 1 where various AC and DC sources and loads are connected to the corresponding DC and AC networks through transformers and converters, respectively. The DC and AC networks are connected together through the bidirectional AC/DC converters which may be transformerless or with transformer.

The PV conversion systems and fuel cell generators are connected to the DC network through DC/DC boosters. Light wind turbine which needs battery as energy buffer can also be connected to DC network. DC loads such as electric vehicles (EVs) and LEDs are connected to the DC network through DC/DC buck converters. Power electronics driven AC motors are connected to DC network through DC/AC converters. DC energy storages such as batteries and super capacitors are connected to the DC network through bidirectional DC/DC converters.

AC power generators such as wind turbine generators and small diesel generators are connected to the AC network. AC energy storages such as flywheels are connected to AC grid through AC/AC converters. AC loads such as AC motors and heaters are connected to the AC network. The three-phase AC network of the

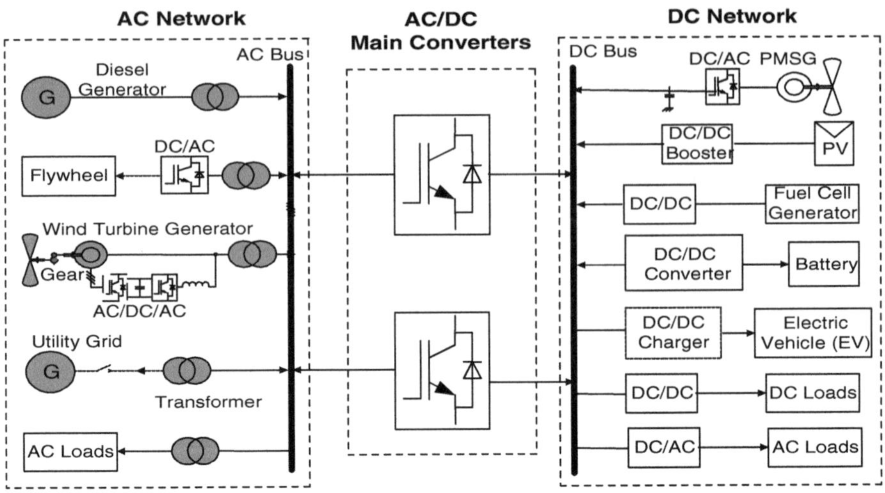

Fig. 1 A typical hybrid AC/DC microgrid configuration

hybrid grid can also be existing in low-voltage distribution network. Hybrid grid can be an isolated grid or connected to the utility grid through a transformer. It should be noted that a transformer may be required for the connection of some AC sources and loads if their output or input voltages are different with the AC network.

The voltage level of the AC grid is 400 V. There is still no standard voltage level for the DC network. The common voltage level currently used in most test systems is 380 V.

3 Modeling and Control of Wind Turbine Generator

Conventional AC power sources are usually synchronous generators and renewable power sources connected to the AC network are usually wind turbine generators. Basic equivalent circuit and operating principle of a synchronous generator can be found in many text books and are not presented in this chapter. The modeling and control of wind turbine generator have been well studied [12–15]. The control and operation of wind turbine generator are presented in this section. The mechanical power output P_m from a WTG can be calculated as

$$P_m = 0.5\,\rho\,\mathrm{A}C_p(\lambda,\,\beta)V_w^3, \tag{1}$$

where ρ is air density, A is rotor swept area, V_w is wind speed and $C_p(\lambda,\,\beta)$ is power coefficient which is the function of tip speed ratio λ and pitch angle β.

The circuit models of a DFIG are essential requirements for WTG control system. The voltage equations of an induction motor in a rotating *d-q* reference frame are as [12]

$$\begin{bmatrix} u_{ds} \\ u_{qs} \\ u_{dr} \\ u_{qr} \end{bmatrix} = \begin{bmatrix} -R_s & 0 & 0 & 0 \\ 0 & -R_s & 0 & 0 \\ 0 & 0 & R_r & 0 \\ 0 & 0 & 0 & R_r \end{bmatrix} \begin{bmatrix} i_{ds} \\ i_{qs} \\ i_{dr} \\ i_{qr} \end{bmatrix} + p \begin{bmatrix} \lambda_{ds} \\ \lambda_{qs} \\ \lambda_{dr} \\ \lambda_{qr} \end{bmatrix} + \begin{bmatrix} -\omega_1 \lambda_{qs} \\ \omega_1 \lambda_{ds} \\ -\omega_2 \lambda_{qr} \\ \omega_2 \lambda_{dr} \end{bmatrix} \tag{2}$$

$$\begin{bmatrix} \lambda_{ds} \\ \lambda_{qs} \\ \lambda_{dr} \\ \lambda_{qr} \end{bmatrix} = \begin{bmatrix} -L_s & 0 & L_m & 0 \\ 0 & -L_s & 0 & L_m \\ -L_m & 0 & L_r & 0 \\ 0 & -L_m & 0 & L_r \end{bmatrix} \begin{bmatrix} i_{ds} \\ i_{qs} \\ i_{dr} \\ i_{qr} \end{bmatrix} \tag{3}$$

The dynamic equation of a DFIG is as:

$$\frac{J}{n_p} \frac{\mathrm{d}\omega_r}{\mathrm{d}t} = T_m - T_{em} \tag{4}$$

$$T_{em} = n_p L_m (i_{qs} i_{dr} - i_{ds} i_{qr}) \tag{5}$$

where the subscripts d, q, s and r denote d-axis, q-axis, stator and rotor, respectively, L represents the inductance, λ is the flux linkage, u and i represent voltage and current, respectively, ω_1 and ω_2 are the angular synchronous speed and slip speed, respectively, ω_r is the rotor speed, $\omega_2 = \omega_1 - \omega_r$, T_m is the mechanical torque, T_{em} is the electromagnetic torque, R_s and L_s are stator resistance and inductance repetitively, R_r and L_r are rotor resistance and inductance repetitively, L_m is the mutual inductance, J is the rotor inertial constant, and n_p is the number of poles.

If the synchronous rotating *d-q* reference is oriented by the stator voltage vector, the d-axis is aligned with the stator voltage vector while the q-axis is aligned with the stator flux reference frame. Therefore, $\lambda_{ds} = 0$ and $\lambda_{qs} = \lambda_s$. The following equations can be obtained in the stator voltage oriented reference frame as [12]:

$$i_{ds} = -\frac{L_m}{L_s} i_{dr} \quad T_{em} = n_p \frac{L_m}{L_s} \lambda_s i_{dr} \quad \sigma = \frac{L_s L_r - L_m}{L_s L_r} \tag{6}$$

$$u_{dr} = R_r i_{dr} + \sigma L_r \frac{\mathrm{d}i_{dr}}{\mathrm{d}t} - (\omega_1 - \omega_r)(L_m i_{qs} + L_r i_{qr}) \tag{7}$$

$$u_{qr} = R_r i_{qr} + \sigma L_r \frac{\mathrm{d}i_{qr}}{\mathrm{d}t} + (\omega_1 - \omega_r)(L_m i_{ds} + L_r i_{dr}) \tag{8}$$

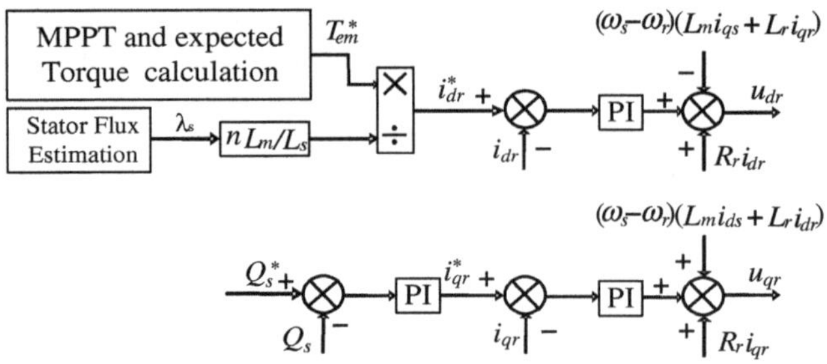

Fig. 2 The DTC control scheme for the rotor side converter

The AC/DC/AC converter of the DFIG is controlled to regulate rotor side current to achieve maximum power point tracking (MPPT), to synchronize with AC grid and to manage the stator side reactive power. Different control schemes such as the direct torque control (DTC) and direct power control (DPC) have been proposed for a DFIG [12–14]. The DTC scheme as shown in Fig. 2 is selected as the control method for the rotor side converter. The rotor rotational speed is obtained through the MPPT algorithm, which is based on the power and speed characteristic of the wind turbine [15]. The rotational speed ω_r and mechanical power P_m are used to calculate the electromagnetic torque T_{em}^*. The d-axis rotor side current reference is determined based on T_{em}^* through stator flux estimation. The rotor side d-q voltages are maintained through controlling the corresponding current with appropriate feed forward voltage compensation.

4 Modeling and Control of Photovoltaic System

One type of DC power sources is PV panels. Individual PV conversion systems have been well studied and the related techniques have been investigated [16–20]. A PV panel is simulated as a current source connected in parallel with a diode and a resistor R_P and in series with resistor R_S as shown in Fig. 3.

The output current I_{PV} of a PV panel can be calculated as [16, 17]

$$I_{\mathrm{PV}} = n_p I_{\mathrm{ph}} - n_p I_{\mathrm{sat}} \times \left[\exp\left(\frac{q}{\mathrm{AkT}}\left(\frac{V_{\mathrm{PV}}}{n_S} + I_{\mathrm{PV}} R_S\right)\right) - 1\right] \quad (9)$$

$$I_{\mathrm{ph}} = (I_{\mathrm{sso}} + k_i(T - T_r)) \times \frac{S}{1000} \quad (10)$$

$$I_{\mathrm{sat}} = I_{rr}\left(\frac{T}{T_r}\right)^3 \exp\left(\left(\frac{qE_{\mathrm{gap}}}{kA}\right) \times \left(\frac{1}{T_r} - \frac{1}{T}\right)\right) \quad (11)$$

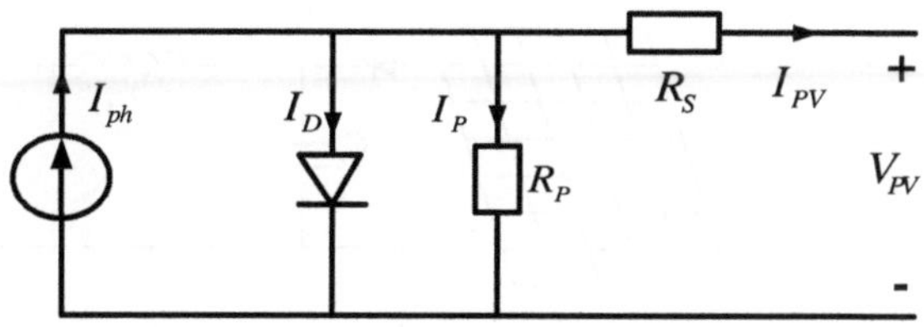

Fig. 3 The circuit model of a PV panel

where V_{PV} is the terminal voltage of PV panel, I_{ph} is the photocurrent, I_{sat} is the saturation current, q is the electron charge, A is the ideality factor, k is the Boltzman constant, I_{sso} is the short circuit current, k_i is the short circuit current temperature coefficient, T_r is the reference temperature, I_{rr} is the reverse saturation current at T_r, E_{gap} is the gap energy, n_p is the number of parallel solar cells, n_s is the number of series solar cells, S is the solar irradiation level and T is the junction temperature.

The output power of a PV panel depends on I-V curve as shown in Fig. 4 where I_{sc} and V_{oc} are the short circuit current and open-circuit voltage, respectively. Because of nonlinear relationship between the output current and the terminal voltage described in Eq. (9), the output power P_{PV} changes with the output current. A Maximum Power Point (MPP) exists at the terminal voltage V_{mp} and the output current I_{mp}. Therefore, a PV panel should be controlled to operate at the MPP.

A PV panel is connected to the DC bus through a DC/DC bucker, booster or buck–boost converter which depends on the terminal voltage of the panel and the DC bus voltage. The layout and control schematic diagram of a basic DC/DC booster for the integration of a PV system is shown in Fig. 5. The PV system is normally controlled to operate in MPPT mode to harness the maximum power. The MPPT techniques including Perturbation & Observe (P&O), Incremental

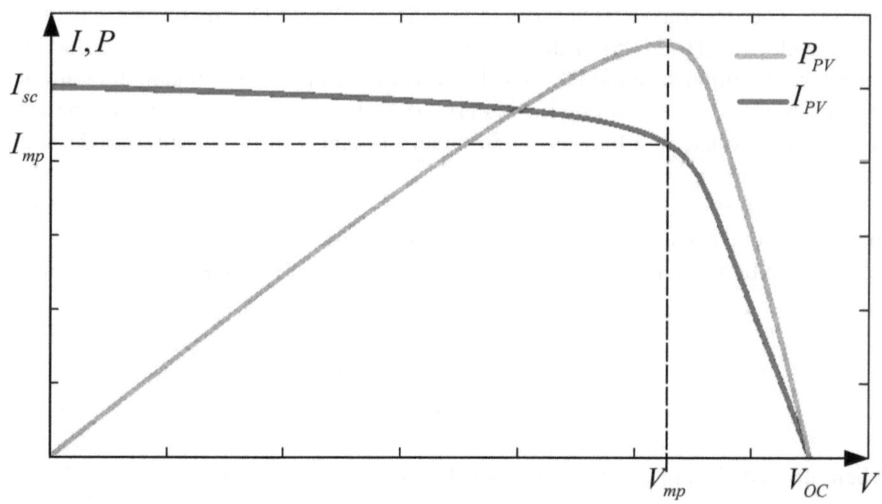

Fig. 4 The *I-V* and *P-V curves* of a PV panel

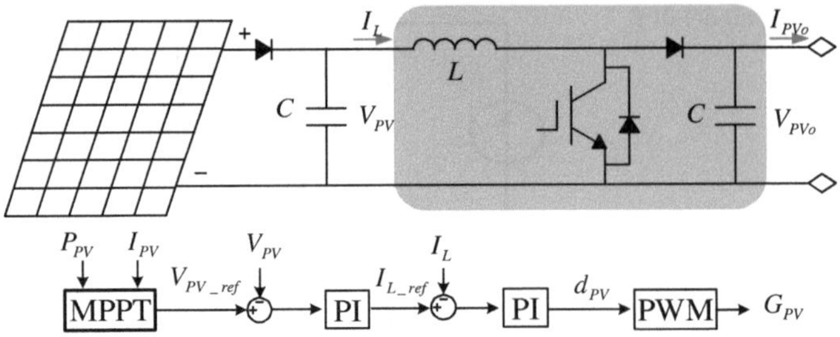

Fig. 5 DC/DC booster converter and the control block diagram for a PV panel

Conductance (IC), fractional open-circuit voltage, etc., have been well developed and introduced in [18–20]. The reference terminal voltage $V_{\mathrm{PV\_ref}}$ is generated in the MPPT function block as shown in the figure and tracked by the conventional double-loop PI controller. In the outer voltage loop, the actual terminal voltage V_{PV} is compared with $V_{\mathrm{PV\_ref}}$ and the error is processed with PI controller to generate the reference inductor current $I_{\mathrm{L\_ref}}$. The inductor current is tracked by a PI controller to generate the duty ratio d_{PV} which is sent to the pulse-width-modulation (PWM) generator to produce the switching signal G_{PV} for generating the maximum power.

5 Modeling and Control of Battery Energy Storage

Battery energy storage system (BESS) is usually designed and connected to the DC bus to maintain power balance between power generation and loads in the DC network. The integration techniques of BESSs to AC and DC microgrids have been well developed [21–27]. A Bidirectional DC/DC converter is used to interface the battery bank output with the DC bus. The layout and control schematic diagram of a DC/DC buck–boost converter for battery control is as shown in Fig. 6. The converter is controlled as a booster when battery operates in discharging mode and a bucker in charging mode.

The upper and lower switches in Fig. 6 are controlled to operate in complementary manner with the certain dead-time to prevent short circuit fault. The relationship between the converter output voltage V_{bo} and the battery output voltage V_b is as

$$V_{\mathrm{bo}} = \begin{cases} \frac{1}{1-d_{\mathrm{bl}}} V_o(\text{discharging}) \\ \frac{1}{d_{\mathrm{bu}}} V_o(\text{charging}) \end{cases} \tag{12}$$

where d_{bl} and d_{bu} are the duty ratio for the lower and upper switches of the battery converter, respectively.

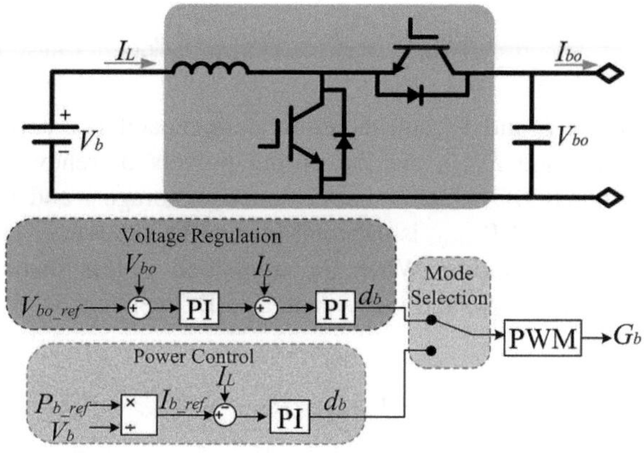

Fig. 6 The circuit and the control block diagram of a bidirectional battery converter

Battery converter can be controlled to operate in both voltage regulation and power control modes [21]. In voltage regulation mode, the reference voltage $V_{\mathrm{bo\_ref}}$ of the battery converter is defined. The conventional double-loop PI control is implemented to track the reference voltage. BESS maintains system power balance autonomously through charge and discharge. The battery output power $P_{\mathrm{b\_ref}}$ is defined in the power control mode. Reference battery output current $I_{\mathrm{b\_ref}}$ is obtained by dividing the power reference with battery terminal voltage. The battery converter duty ratio is generated in the current control loop with the PI controller. Upon the determination of the operating mode in the mode selection block, the duty ratio is sent to the PWM generator. The constraints of BESS operation include the limitations of battery charge/discharge current and the rated energy capacity. Over-current charge/discharge degrades battery lifetime. Therefore, battery current is monitored in real time. In case of battery current exceeding the predefined threshold value, converter power/current control will be activated. The maximum allowable charge/discharge current is set as the current reference. The actual energy stored in battery is normally indicated by the State of Charge (SoC) or the Depth of Discharge (DoD). When the SoC exceeds the upper or lower limitation, over-charge/discharge damage is induced. When the SoC reaches the maximum value, battery is controlled to operate in only the discharging mode, and vice versa.

6 Operation and Control of DC Network

Operation and control techniques of AC power systems are very mature and will not be discussed in this chapter. Reliable and economic operation of the DC network depends on the coordinated control of multiple sources, loads, and energy storages. The control techniques for the DC microgrid have been proposed [6, 7, 8, 9, 10, 11, 28]. DC bus voltage is an important indicator of power balance of the DC network. The relationship between DC bus voltage and net power of DC network at time t is as

$$CV_{dc} = \frac{dV_{dc}}{dt} = (P_{RESs} + P_{DGs} + P_{BESs} + P_{BIC} - P_{Loads}), \quad (13)$$

where C and V_{dc} are the equivalent capacitance and voltage of the DC bus, P_{RESs}, $P_{DGs,}$ and P_{BESs} are the output powers of renewable energy sources, other distributed DC sources, battery energy storages, and bidirectional converter, respectively, and P_{Loads} is the total load of DC network. The sign of a P_{BES} indicates the operating mode. When the total load is less than the total generation, DC bus voltage drops, and vice versa. The DC network is in the steady state, the net power is as

$$P_{RESs} + P_{DGs} + P_{BESs} + P_{BIC} - P_{Loads} = 0 \quad (14)$$

Keeping the real-time power balance of the DC bus is a complicated control problem when considering multiple objectives of system operation including maximizing renewable energy harvest, optimizing usages of BESSs and maintaining stable DC bus voltage within its limits under loads and resources variations. A three-level hierarchical level (HL) control of HLI, HLII, and HLIII is proposed to achieve reliability and economic operation of the DC network.

In the hierarchical control, the DC bus voltage is divided into five regions using the four predefined voltage thresholds whose values are determined by system operation requirements. The relationship among the threshold values are as.

$$V_{L2} < V_{L1} < V_{dcn} < V_{H1} < V_{H2}, \quad (15)$$

where V_{dcn} is the nominal DC bus voltage, V_{L2} and V_{H2} are the low and up limits of the DC network, respectively, V_{H1} and V_{L1} are threshold values for activating the battery charging and discharging, respectively. It should be noted that the threshold value varies based on the nominal voltage and operation requirements of a particular system. V_{H2} and V_{H1} are usually set to be 10 and 5 % above the nominal voltage, respectively, while V_{L1} and V_{L2} are 5 and 10 % below nominal voltage, respectively. For stable operation of the DC network, at least one of the DC sources should be controlled to regulate bus voltage (slack terminal). The five voltage regions are as follows:

Region 1: $V_{L1} \leq V_{dc} \leq V_{H1}$
Region 2: $V_{H1} < V_{dc} \leq V_{H2}$
Region 3: $V_{L2} \leq V_{dc} \leq V_{L1}$
Region 4: $V_{H2} \leq V_{dc}$
Region 5: $V_{L2} \geq V_{dc}$

6.1 Operation and Control of DC Network in HLI

The main objective of HLI control is to harvest the maximum powers from renewable energy sources and maintain basic operating reliability. The control schemes for different operation regions are as follows.

Region 1: To prevent frequent battery charging/discharging due to random variation of loads or renewable sources, Region 1 is the operation band in which all BESSs operate in the idle mode. PV converters work in MPPT mode to harness the maximum renewable energy to supply load. Power is balanced between power sources and loads through voltage regulation. Due to the lack of slack terminal, the bus voltage is allowed to change within the Region 1 according to power variation of loads and sources.

Region 2: In this region, there is power generation surplus from renewable sources. PV converters are controlled to operate in MPPT mode and the BESSs are activated to store the surplus power. All other nonrenewable power sources are in idle mode. Charging power sharing among BESSs is implemented by the droop control. The charging current of *i*th BESS is determined by the linear droop as

$$V_{\text{boi}}^{\text{ref}} = V_{\text{H1}} - m_{\text{bi}} I_{\text{boi}}, \text{ (charging)}, \tag{16}$$

where $V_{\text{boi}}^{\text{ref}}$, I_{boi} and m_{bi} are the reference voltage, the output current of *i*th battery converter and the droop coefficient of *i*th BESS, respectively.

Region 3: In this region, the DC bus voltage is low because power generation is less than load. PV converters are controlled to operate in the MPPT mode and BESSs are in discharging mode to compensate the power shortage. When BESSs reach the maximum discharging rate, to prevent bus voltage from collapsing, nonrenewable power sources change from the idle mode to the voltage regulation mode at V_{L2}. The discharging current of *i*th BESS is determined by the linear droop as

$$V_{\text{boi}}^{\text{ref}} = V_{\text{L1}} - m_{\text{bi}} I_{\text{boi}}, \text{ (discharging)} \tag{17}$$

6.2 Operation and Control of DC Network in HLII

The limitation of the HLI control is that all the resources are controlled by local voltage signal. The utilization of the resources is not optimal. This can be improved by real-time information sharing among converters through communication links and energy management system in HLII control. Communication link generally brings full observability over DC microgrid including the real-time bus voltage, power flow, and operation status of converters, thus system optimal operation in term of reliability and cost could be achieved through coordination control on system elements. Although the effectiveness of system operation depends heavily on the communication quality in HLII, the system reliability would not be degraded due to HLI as the backup. Once failure happens in the communication links, all elements including PV modules and battery storages can retain the system stability under HLI control with the cost of losing global optimization. The control strategy for Region 1 in HLII is the same as the one in HLI. The PV modules operate in MPPT mode and BESSs are idle. The main objective of HLII control is to solve the power sharing problems among PV modules and BESSs when there is PV power surplus or shortage.

6.3 Operation and Control of DC Network in HLIII

Over or under voltage may occur due to sudden loss of loads or PV modules in system operation, which may damage system components or lead to malfunction of system control. To protect those abnormal conditions, HLIII control is therefore designed to maintain the DC bus voltage within allowable band when abnormal system conditions occur.

Region 4: When load demand is low, renewable generation is high, and all BESSs are fully charged, the DC bus voltage may be over the limitation. In this case, some of PV converters have to be switched from the MPPT mode to voltage regulation mode to maintain system normal operation.

Region 5: When all renewable and nonrenewable sources and BESSs are operate in their maximum capacity and still cannot supply the load, the DC bus voltage may be under the limitation. In this case, some of DC load have to be curtailed.

7 Operation and Control of Bidirectional DC/AC Converter

The hybrid grid can operate in two modes [29]. In grid-connected mode, the main converter is to provide stable DC bus voltage and required reactive power and to exchange power between the AC and DC buses. The PV converters and WTG are controlled to provide the maximum power. When the output power of the DC sources is greater than the DC loads, the converter should be controlled as an inverter and injects power from DC to AC side. When the total power generation is less than the total load at DC side, the main converter transfers power from the AC to DC side. When the total power generation is greater than the total load in the hybrid grid, it will produce power to the utility grid. Otherwise, the hybrid grid will absorb power from the utility grid. In grid-tied mode, the battery converter is not very important in system operation because power is balanced by the utility grid.

In autonomous mode, battery plays a very important role for both power balance and voltage stability. Control objectives for various converters can be centrally dispatched by energy management system (EMS) or decentralized. The DC bus voltage is maintained stable by battery converters or boost converters according to different operating conditions. The main converter is controlled to provide a stable and high quality AC bus voltage. Both PV and WTG can operate in MPPT or off-MPPT mode based on system operating requirements.

There are PV and storage converters in the DC network, the DFIG back-to-back converters in the AC network and the main converter interconnecting AC/DC networks. Those converters have to be coordinately controlled with the utility grid to supply an uninterrupted, high efficiency, and high quality power to variable DC and AC loads under variable solar irradiation and wind speed when the hybrid grid operates in both isolated and grid-tied modes. The control algorithms for PV, BESS, and WTG converters have been introduced in previous sections. The control and operation of the bidirectional main converter are presented in this section.

There are the centralized and decentralized control methods for operation of hybrid microgrids. Decentralized control for multiple converters without communication links provides a good solution to ride through communication malfunction and therefore enhance the system reliability. The fully decentralized control for an autonomous hybrid microgrid is not straightforward if power exchange between two sub-grids is required. In this section, a fully decentralized control for power management in individual sub-grid and throughout the entire hybrid system is presented by means of local power sharing in individual AC and DC networks and global power sharing between two networks.

7.1 Local Power Sharing in the AC Network

WTGs in the AC network are controlled to operate in MPPT mode. Power balance between generation and load is maintained by controlling the outputs from the conventional DGs (CDGs). Power sharing among the CDGs can be realized by the droop control method [30]. The *P-f* droop curve for *x*th CDG unit can be expressed as

$$f = f^* + m_x P_{ac\_x} \tag{18}$$

$$m_x = \frac{f_{\min} - f_{\max}}{P_{ac\_x}^{\max}} \tag{19}$$

where f^* is the reference frequency of the AC network, P_{ac_x} is power output, m_x is the droop coefficient, $f_{\min}$ and $f_{\max}$ are the minimum and maximum allowable frequency respectively, and $P_{ac\_x}^{\max}$ is the maximum active power. Based on the same system frequency for all DGs, the power sharing among *u* CDGs can be obtained as

$$\frac{P_{ac\_1}}{P_{ac\_1}^{max}} = \frac{P_{ac\_2}}{P_{ac\_2}^{max}} = \cdots = \frac{P_{ac\_u}}{P_{ac\_u}^{max}} \tag{20}$$

7.2 Local Power Sharing in the DC Network

As discussed in the previous section, all the renewable sources are controlled to operate in MPPT mode. The DC bus voltage is maintained by charging or discharging of BESSs. The droop control is applied for power sharing among BESSs. The droop curve of yth BESS in the DC network is as

$$V_{dc\_y} = V_{dc}^{*} + d_y P_{dc\_y} \tag{21}$$

$$d_y = \frac{V_{dc}^{min} - V_{dc}^{max}}{P_{dc\_y}^{max}} \tag{22}$$

where V_{dc}^{*} is the reference DC bus voltage and configured as the maximum allowable voltage, V_{dc_y} is the terminal voltage of BESS and d_y is the negative droop coefficient, V_{dc}^{min} and V_{dc}^{max} are the minimum and maximum allowable DC bus voltage in the DC network, respectively, and $P_{ac\_y}^{max}$ the is maximum output power for yth BESS.

The terminal voltages of BESSs are slightly different with bus voltage V_{dc} due to the different voltage drop in the cable. The voltage difference would lead to the inaccurate power sharing among BESSs. To solve this problem in DC network, a modified droop equation is developed as

$$V_{dc\_y} = V_{dc}^{*} + d_y P_{dc\_y} + i_{dc\_y} Z_{dc\_y}, \tag{23}$$

where i_{dc_y} and Z_{dc_y} are the output current and cable impedance. The power sharing equation among v BESSs can be obtained as

$$\frac{P_{dc\_1}}{P_{dc\_1}^{max}} = \frac{P_{dc\_2}}{P_{dc\_2}^{max}} = \cdots = \frac{P_{dc\_v}}{P_{dc\_v}^{max}} \tag{24}$$

7.3 Global Power Sharing Among AC and DC Networks

Based on DC bus voltage and frequency of AC network, the control algorithm for power exchange between two networks is required. Because of the complexities of control and power sharing in two networks, a normalization process is proposed to determine the control parameters of bidirectional converter. The two separate droop characteristics are normalized and combined to determine the power transfer between AC/DC networks [29].

Considering u CDGs in the AC network, the combined AC droop characteristics of the AC network is as

$$f = f^* + M \sum_{x=1}^{u} P_{\mathrm{ac\_}x} \tag{25}$$

$$M = 1 \Big/ \sum_{x=1}^{u} \frac{1}{m_x} \tag{26}$$

where $\sum_{x=1}^{u} P_{\mathrm{ac\_}x}$ is the total active power generated from u CDG units. M is the combined droop coefficient of the AC network.

Considering all BESSs in the DC network, the combined DC droop characteristics of DC network is as

$$V_{\mathrm{dc}} = V_{\mathrm{dc}}^* + D \sum_{y=1}^{v} P_{\mathrm{dc\_}y} \tag{27}$$

$$D = 1 \Big/ \sum_{y=1}^{v} \frac{1}{d_y} \tag{28}$$

where $\sum_{y=1}^{v} P_{\mathrm{dc\_}y}$ is the total power generated from all v BESS units in the DC network. D is the combined droop coefficient of the DC network.

Based on the combined droop equations of AC and DC networks, a normalization method is used to unify two control variables f and V_{dc} with different dimensions. The two combined droops in (25) and (27) are normalized as

$$(\gamma)' = \begin{cases} \frac{\gamma - \gamma_n}{\gamma_{\max} - \gamma_n}, & \text{for} \quad \gamma > \gamma_n \\ \frac{\gamma_n - \gamma}{\gamma_{\min} - \gamma_n}, & \text{for} \quad \gamma < \gamma_n \end{cases} \tag{29}$$

where γ represents f or V_{dc}, $(\gamma)'$ is normalized γ, γ_n is the nominal value of γ, and $\gamma_{\max}$, and $\gamma_{\min}$ are the maximum and minimum values of γ, respectively. For $(\gamma_{\max} - \gamma_n) = (\gamma_n - \gamma_{\min})$, $(\gamma)'$ can be written as

$$(\gamma)' = \frac{\gamma - 0.5(\gamma_{\max} + \gamma_{\min})}{0.5(\gamma_{\max} - \gamma_{\min})} \tag{30}$$

The normalized value $(\gamma)'$ varies between −1 and 1. Substituting (30) into (25) and (27), the normalized droop characteristics for the respective AC and DC networks are as

$$(f)' = 1 + \frac{M}{0.5(f_{\max} - f_{\min})} \sum_{x=1}^{u} P_{\mathrm{ac\_}x} \tag{31}$$

$$(V_{\mathrm{dc}})' = 1 + \frac{D}{0.5\left(V_{\mathrm{dc}}^{\max} - V_{\mathrm{dc}}^{\min}\right)} \sum_{y=1}^{v} P_{\mathrm{dc\_}y} \tag{32}$$

Considering the definitions of m_x and d_y, (31) and (32) can be manipulated into (33) and (34), respectively.

$$(f)' = 1 - \frac{1}{0.5 \sum_{x=1}^{u} P_{\mathrm{dc\_}x}^{\max}} \sum_{x=1}^{u} P_{\mathrm{dc\_}x}^{\max} \tag{33}$$

$$(V_{\mathrm{dc}})' = 1 \frac{1}{0.5 \sum_{y=1}^{v} P_{\mathrm{dc\_}y}^{\max}} \sum_{y=1}^{v} P_{\mathrm{dc\_}y}^{\max} \tag{34}$$

The combined droop characteristics of two subnetworks have same dimensions on x and y axis and can be placed in the same reference frame with common vertical and horizontal axis. Global power sharing can then be inferred from the local power sharing in AC or DC network, whose f/V_{dc} are inherently unique to ensure proportional power sharing among all the units in both AC and DC networks. The attempt is to employ a PI controller to equalize $(f)'$ and $(V_{\mathrm{dc}})'$ by exchanging power P between the AC and DC networks as

$$P = \left[(f)' - (V_{\mathrm{dc}})'\right]\left(k_p + k_i/s\right) \tag{35}$$

where k_p and k_i are controller gains, and P is the power flows from AC to DC network. If $(f)'$ is larger than $(V_{\mathrm{dc}})'$, power is transferred from AC to DC network. If $(f)'$ is less than $(V_{\mathrm{dc}})'$, power is transferred from DC to AC network. If $(f)'$ equals to $(V_{\mathrm{dc}})'$, there is no power transfer. The control scheme based on (35) ensure that all DGs in AC or DC networks can share the total load of hybrid grid in proportion to their maximum powers throughout the hybrid AC/DC system.

8 Summary

This chapter presents a new configuration for future power systems which is the hybrid AC/DC gird for high efficient connection of the inherent AC and DC sources and loads. Conventional AC and DC grids are interconnected together through the bidirectional AC/DC converter. The component model has been introduced. The control and operation of individual sources and energy storages are presented. The coordination control and power sharing techniques are also introduced.

References

1. Wang P, Goel L, Liu X, Choo F (2013) Harmonizing AC and DC: a hybrid AC/DC future grid solution. IEEE Power Energy Mag 11(3):76–83
2. Zoka Y, Sasaki H, Yorino N, Kawahara K, Liu CC (2004) An interaction problem of distributed generators installed in a MicroGrid. In: Proceedings of the IEEE electric utility deregulation, restructuring and power technologies, vol 2, April 2004, pp 795–799
3. Lasseter RH, Paigi P (2004) Microgrid: a conceptual solution. In: Proceedings of the IEEE 35th PESC, vol 6, June 2004, pp 4285–4290
4. Sao CK, Lehn PW (2008) Control and power management of converter fed microgrids. IEEE Trans Power Syst 23(3):1088–1098
5. Logenthiran T, Srinivasan D, Wong D (2008) Multi-agent coordination for DER in Micro Grid. In: Proceedings of the IEEE international conference on sustainable energy technologies, Nov 2008, pp 77–82
6. Baran ME, Mahajan NR (2003) DC distribution for industrial systems: Opportunities and challenges. IEEE Trans Ind Appl 39(6):1596–1601
7. Ito Y, Yang Z, Akagi H (2004) DC micro-grid based distribution power generation system. In: Proceedings of the IEEE international power electronics and motion control conference, vol 3, Aug 2004, pp 1740–1745
8. Sannino A, Postiglione G, Bollen MHJ (2003) Feasibility of a DC network for commercial facilities. IEEE Trans Ind Appl 39(5):1409–1507
9. Hammerstrom DJ (2007) AC versus DC distribution systems-did we get it right?. In Proceedings of the IEEE power engineering society general meeting, June 2007, pp 1–5
10. Salomonsson D, Sannino A (2007) Low-voltage DC distribution system for commercial power systems with sensitive electronic loads. IEEE Trans Power Deliv 22(3):1620–1627
11. Xiao J, Wang P (2013) Multiple modes control of household DC microgrid with integration of various renewable energy sources. In: Industrial electronics society, IECON 2013—39th annual conference of the IEEE, 2013 pp 1773–1778
12. Zhi W, Xu L (2007) Direct power control of DFIG with constant switching frequency and improved transient performance. IEEE Trans Energy Convers 22(1):110–118
13. Arnalte S, Burgos JC, Rodriguez-amenedo JL (2002) Direct torque control of a doubly-fed induction generator for variable speed wind turbines. Electr Power Compon Syst 30(2):199–216
14. Kim WS, Jou ST, Lee KB, Watkins S (2008) Direct power control of a doubly fed induction generator with a fixed switching frequency. In: Proceedings of the IEEE industry applications society annual meeting, Oct 2008, pp 1–9
15. Koutroulis E, Kalaitzakis K (2006) Design of a maximum power tracking system for wind-energy-conversion applications. IEEE Trans Industr Electron 53(2):486–494
16. Ropp ME, Gonzalez S (2009) Development of a MATLAB/simulink model of a single-phase grid-connected photovoltaic system. IEEE Trans Energy Convers 24(1):195–202
17. Chao KH, Li CJ, Ho SH (2008) Modeling and fault simulation of photovoltaic generation systems using circuit-based model. In: Proceedings of the IEEE international conference on sustainable energy technologies, Nov 2008, pp 290–294
18. Esram T, Chapman PL (2007) Comparison of photovoltaic array maximum power point tracking techniques. IEEE Trans Energy Convers 22:439–449

19. Sung-Jun K, Jae-Sub K, Jung-Sik C, Mi-Geum J, Ju-Hui M, Jin-Gook L et al (2011) A novel MPPT control of photovoltaic system using FLC algorithm. In: 2011 11th international conference on Control, automation and systems (ICCAS), 2011, pp 434–439
20. Femia N, Petrone G, Spagnuolo G, Vitelli M (2005) Optimization of perturb and observe maximum power point tracking method. IEEE Trans Power Electron 20:963–973
21. Wang P, Xiao J, Setyawan L, Jin C, Hoong CF (2014) Hierarchical control of active hybrid energy storage system (HESS) in DC microgrids. In: 2014 IEEE 9th conference on industrial electronics and applications (ICIEA), 2014, pp 569–574
22. Anbuky AH, Pascoe PE (2000) VRLA battery state-of-charge estimation in telecommunication power systems. IEEE Trans Ind Electron 47(3):565–573
23. Kutluay K, Cadirci Y, Ozkazanc YS, Cadirci I (2005) A new online state-of-charge estimation and monitoring system for sealed lead-acid batteries in Telecommunication power supplies. IEEE Trans Ind Electron 52(5):1315–1327
24. Hill CA, Such MC, Dongmei C, Gonzalez J, Grady WM (2012) Battery energy storage for enabling integration of distributed solar power generation. IEEE Trans Smart Grid 3:850–857
25. Etxeberria A, Vechiu I, Baudoin S, Camblong H, Vinassa J (2012) Control of a hybrid energy storage system using a three level neutral point clamped converter. In: IECON 2012—38th annual conference on IEEE industrial electronics society, 2012, pp 3400–3405
26. Hao Q, Jianhui Z, Jih-Sheng L, Wensong Y (2011) A high-efficiency grid-tie battery energy storage system. IEEE Trans Power Electron 26:886–896
27. Guerrero JM, Lijun H, Uceda J (2008) Control of distributed uninterruptible power supply systems. IEEE Trans Ind Electron 55(8):2845–2859
28. Xiong L, Peng W, Poh Chiang L (2011) A hybrid AC/DC microgrid and its coordination control. IEEE Trans Smart Grid 2:278–286
29. Wang P, Jin C, Zhu D, Tang Y, Loh P, Choo F (2014) Distributed control for autonomous operation of a three-port AC/DC/DS hybrid microgrid. In: Accepted for publication in IEEE transactions on industrial electronics, 2014
30. Jin C, Wang P, Xiao J, Tang Y, Choo F (2014) Implementation of hierarchical control in DC microgrids. IEEE Trans Industr Electron 61(8):4033–4042

Chapter 3
Low Voltage Ride Through of Wind Energy Systems

Tarafdar Hagh Mehrdad and Kashem Mohammad Muttaqi

Abstract This chapter titled, "Low Voltage Ride Through of Wind Energy Systems" focuses on the importance of low voltage ride through (LVRT) of wind turbines. The investigation of some selected grid codes has revealed that LVRT has different requirements in different grid codes. The results of the investigation have found that higher penetration of wind energy in grid results in stricter LVRT requirements. On the other hand, increasing the wind energy penetration in grid is an ongoing process; so it will be mandatory to increase the LVRT performance of existing wind turbines as well as newly installed units. This chapter has presented some key power circuit topology based and control strategy based LVRT improvement methods. While power circuit topology based strategies are usually more efficient than control strategy based methods, they need extra power electronic-based electrical circuits which might be expensive or bulky and they may affect the normal operation of wind turbines, too. This chapter has also discussed the problems of resynchronization of wind turbines. It is shown that without sufficient LVRT capability, wind turbines would be disconnected from grid during faults and resynchronization of them would be a time-consuming process which may impose large transients to the grid. A resynchronization scheme for automatic reconnection of wind turbines has been presented and validated through simulation.

T.H. Mehrdad
Faculty of Electrical and Computer Engineering, University of Tabriz, Tabriz, Iran
e-mail: tarafdar@tabrizu.ac.ir

K.M. Muttaqi (✉)
School of Electrical, Computer and Telecommunications Engineering,
Australian Power Quality and Reliability Centre, University of Wollongong,
Wollongong, NSW, Australia
e-mail: kashem@uow.edu.au

N.R. Karki et al. (eds.), *Sustainable Power Systems*, Reliable and Sustainable Electric Power and Energy Systems Management,
DOI 10.1007/978-981-10-2230-2_3

1 Introduction

Among renewable energies, onshore wind power has already some examples of unsubsidized competitive with conventional energy. By 2015, it is expected that energy cost from onshore wind turbines will reach to 8–10 US cents/kWh and it will achieve parity with fossil fuel plants in wholesale market.

While the investment in wind energy is growing rapidly, there are still some questions which should be answered from power system stability point of view, some of which are as follows:

- What are the related issues with high penetration of wind energy in power system?
- How is it possible to modify the performance of existing wind turbines to cope with new power system requirements?

The literature review shows that there are two main issues related to wind turbines' effect on power system stability which are:

- Power system dynamic stability, and
- Power system transient stability.

Power system dynamic stability generally concerned to small disturbances lasting for about 10–30 s. Main dynamic stability issues related to wind turbines are

- Frequency response including the performance of fixed and variable speed wind turbines,
- Real-time pricing including active load management, and
- Energy storage application to improve the dynamic stability of wind turbines including application of super-capacitor and Superconductor Magnetic Energy storage (SMES).

Power system transient stability means the ability of power system to reach a stable condition following a large disturbance like sudden connection or disconnection of heavy loads, switching operations, and faults. Main transient stability issues related to wind turbines are

- Reactive power support during fault, and
- Low Voltage Ride Through (LVRT) of Wind Turbines.

This chapter will focus on low voltage ride through (LVRT) of wind turbines. Part 2 explains three grid codes including ENTSO-E network code (known as harmonized Europe network code), Australia network connection rules (code) and Iranian grid code from LVRT point of view. ENTSO-E is the first practical example of a harmonized code between utilities with different priorities from LVRT point of view. Australia network connection rules and Iran grid codes are good examples of a progressed and developing countries' grid codes, respectively. Last versions of all three grid codes are approved in 2014. Part 3 explains the LVRT and its improvement techniques. This part also explains power circuit topology and control strategy based solutions for LVRT improvement. Part 4 describes the problems related to resynchronization and restoration of wind turbines after major events.

2 Low Voltage Ride Through (LVRT) in Grid Codes

A grid code is a technical specification which determines the necessary parameters of a generating unit for its connection to a public electric network. Grid code is essential for cooperation between turbine manufacturers, wind farm operators, service providers, certification bodies, and engineering companies. The grid codes differ between countries both in the grid code structure and in the technical requirements. The requirements are also subject to continuous revision and changes. Usually, the requirements concern the following topics:

- Active power control
- Frequency tolerance/frequency control
- Reactive power/Voltage regulation
- Communication and controllability
- Tests and documentation including model requirements
- Dynamic grid support, and
- LVRT capability

The extent and duration to which the voltage can fall before the wind turbine is disconnected from the grid is commonly called low voltage ride through (LVRT) or fault ride through (FRT) capability [1]. LVRT and FRT are usually used interchangeably. LVRT is a technical requirement for keeping the wind turbine connected to grid during faults so that they can contribute to the stability of power system. This is contrary to the traditional approach, where the power plants were not required to stay connected to power system during faults. Therefore, LVRT requirement is a key technical issue in the Grid Codes.

Due to increase in the number and capacity of renewable energies, many countries have imposed the LVRT requirement regarding the renewable energy penetration level and local utility practices and knowledge. The LVRT requirement could be very different in different grid codes. In this section, LVRT requirement in European Union (EU) is compared with that in Australia and Iran grid codes (as two non-Europe countries) with different levels of renewable energy penetration and technical issues.

The EU has the highest penetration of wind power with a target of 14–17 % of the EU's electricity production by wind power by 2020. In 2009, European Network of Transmission System Operators for Electricity (ENTSO-E) agreed with European Regulators' Group for Electricity and Gas (ERGEG) and the European Commission (EC) to launch a "pilot code project," aimed at harmonizing grid code requirements for wind generators throughout Europe. This harmonized grid connection requirement is recently approved by Agency for the Cooperation of Energy Regulators (ACER) of Europe on 21 March 2014 [2]. The EU-code is based on the following subdivisions:

- Synchronous area including: Continental Europe, Nordic, Great Britain, Ireland, and Baltic.
- Power Generating Module (PGM) type including: types A, B, C, and D. Table 1 shows connection point voltage and maximum capacity thresholds for different types of PGM. The power rating threshold values should be defined by each relevant Transmission System Operator (TSO) but it cannot be more than the maximum capacity values given in Table 1. Each PGM type should provide the related automated dynamic response and regulations, too. The defined power rating thresholds are in increasing order from type A to type D. For example, if the power rating threshold for type A—PGM has been selected to be 0.001 MW by the relevant TSO, it should be more than 0.001 MW but less than 1 MW. This is because 1 MW is the power rating threshold for type B-PGM.
- Module types including: Synchronous Power Generating Module (SPGM) and Power Park module (PPM). A PPM is connected to the network non-synchronously or through power electronics.

The EU-code does not require special considerations for FRT of generator type A.

Table 2 and the diagram represented by Fig. 1 shows the LVRT requirements in EU-code. In the above-mentioned table and figure, voltage, and time are given in per-unit (pu) and in second (s), respectively. The Uret is the retained voltage at the connection point during fault; Uclear is the voltage recovery at the fault clearing

Table 1 Thresholds for different types of PGM

Generator type	A	B	C	D
Connection point voltage (kV)	<110	<110	<110	≥ 110
Maximum capacity (MW)	0.0008<	1	50	75

Table 2 The LVRT requirements in EU-code

Gen. type	B or C		D	
PGM	SPGM	PPM	SPGM	PPM
U_{ret}	0.05–0.3	0.05–0.15	0	0
t_{Clear}	0.14–0.25			
U_{Clear}	0.7–0.9	U_{ret}–0.15	0.25	U_{ret}
t_{rec1}	t_{Clear}		t_{Clear}–0.45	t_{Clear}
U_{rec1}	U_{Clear}		0.5–0.7	U_{Clear}
t_{rec2}	t_{rec1}–0.7	t_{rec1}	t_{rec1}–0.7	t_{rec1}
U_{rec2}	0.85–0.9 and $\geq U_{Clear}$	0.85	0.85–0.9	0.85
t_{rec3}	1.5–3.0		t_{rec2}–1.5	–3.0

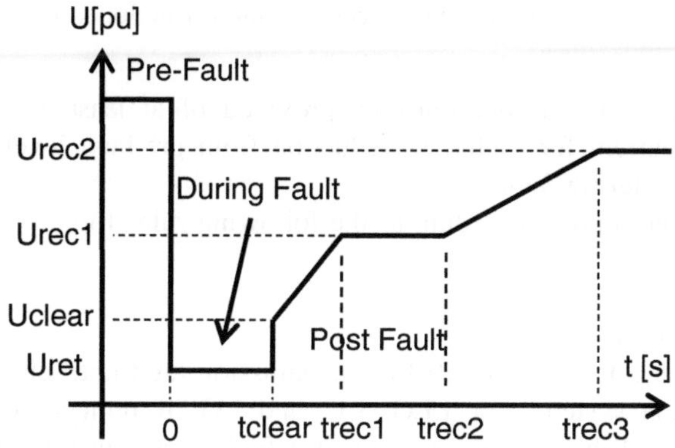

Fig. 1 LVRT diagram of EU-code

time tclear; U_{rec1}, and U_{rec2} specify certain points of lower limits of voltage recovery after fault clearance; and t_{rec1}, t_{rec2}, and t_{rec3} are the times at different voltage recovery.

Australia is a country with a high growth of renewable energy. Table 3 shows some recent figures in Australian electricity generation [3]. Australia has published the last version of its connection code on 1 July 2014 [4].

Australia-code has divided the power generating systems into three categories including: Automatic access standard, minimum access standard and negotiated access standard. These divisions are not based on rating of generating system or unit. A generating unit will be categorized in one of the mentioned access standards if it meets all the standard requirements of that access. These access codes explain LVRT of different Line to line (LL), double line to ground (LLG), or Line to Ground (LG) faults if applicable.

Table 3 Australia renewable energy generation

	2012–2013		Average annual growth	
	GWh	Share %	2012–2013 %	10 years %
Renewable energy	32,566	13.1	26.2	6.2
Bioenergy	3151	1.3	3.5	6.4
Hydro	18,270	7.3	29.7	1.3
Geothermal	1	0.0	0.0	0.0
Wind	7328	2.9	19.9	29.7
Solar PV	3817	1.5	49.2	56.4

The LVRT of Australia automatic access standard includes the following main items:

- Capacitive reactive current must be provided of at least 4 % of maximum continuous current for each 1 % reduction from pre-fault level of connection point voltage during fault.
- Remain uninterrupted according to the following rules for:

1. Transmission faults:
   For three-phase Line to Line (LLL) transmission fault until cleared by breaker fail protection system. For LLG, LL, and LG, if protection system is not installed, uses Table 4. For example, if a fault occurs in a point with voltage level equal or greater than the nominal voltage Vn = 400 kV (as shown in the second row of the table), the electrical generation units which are connected to the utility should remain connected at least for 175 ms. The other rows of Table 4 can be explained in similar manner. If none is specified, it will remain connected to the grid for at least 430 ms:
2. Distribution faults:
   For LLL, LLG, LL, and LG until cleared by breaker fail protection system. If protection system is not installed, the greater of 430 ms (which is the longest time expected) is to be taken for all relevant primary protection system to clear the fault.

The LVRT of Australia minimum automatic access standard includes the following main items:

- Remain uninterrupted for transmission and distribution LLG, LL, and LG faults for the longest time expected to be taken for all relevant primary protection systems to clear the fault unless Australian Energy Market Operator (AEMO) and the Network Service Provider (NSP) agree that total reduction of active power (P) is less than 100 MW and has no adverse impact on the quality or security of the system.
- Deliver P and supply or absorb reactive power (Q) to ensure the generating system response to voltage disturbances conditions, after LLG, LL, and LG faults.

Table 4 Voltage at fault location versus time

Vn at fault location (kV)	Time (ms)
$Vn \geq 400$	175
$250 \leq Vn < 400$	250
$100 < Vn < 250$	430
$Vn \leq 100$	As necessary to prevent plant and meet stability

The LVRT of Australia negotiated access standard includes the following main items:

The plant must not cause other plants or loads to trip as a result of an event. The AEMO and NSP must take into account the expected performance of existing networks, generating plants, projects, controls, and protection systems in addition to the expected range of power system operating conditions.

Iran is producing 0.5 % of its total energy by wind turbines with high average growth of energy production during past years to achieve 5 % of total electricity production by 2020. Iran has published the grid codes for renewable energies in February 2014. As wind power has not penetrated considerably in electrical energy production, there would not be necessary for introducing categorized and detailed LVRT rules. This could promote expansion of wind energy production, too. Iran grid code requires wind turbines to "remain connected until cleared by breaker fail protection system."

Comparison of European Union (EU) with Australia and Iran grid codes as two non-Europe countries with different levels of renewable energy penetration and technical issues shows that by increasing the penetration of wind turbines in power system, stricter LVRT requirements are proposed to be in action by the utilities.

3 Technical Solutions to Improve LVRT in Wind Turbines

Grid faults cause a voltage dip at the connection point of the wind turbine. The type of wind turbine determines its principle of operation during fault. For example, the dip in the grid voltage will result in an increase of the current in the stator windings of the conventional double fed induction generator (DFIG)-based wind turbine and this current will also induces high currents in rotor side which may damage the rotor circuit, the power electronic converters, or DC link capacitor due to its voltage rise. The operation of permanent magnet synchronous generator (PMSG)-based wind turbine which connects to the grid through a back to back converter differs from DFIG under fault condition. In fault condition, the surplus power through the generator-side converter starts to charge the DC link capacitor and cause over-voltage in the DC link of PMSG. In this case, the grid-side converter cannot deliver the energy of DC link to the utility.

There are two main procedures to improve the LVRT of wind turbines which are:

- Power circuit topology based solutions, and
- Control strategy based solutions.

Both the mentioned solutions for improving the LVRT of wind turbines are explained in following subsections.

3.1 Power Circuit Topology Based LVRT Improvement

Power circuit topology based solutions are based on adding a compensator or electrical element to conventional power circuit topology of wind turbine in such a way that the fault current or DC link voltage rise during fault could be limited. This methodology can be subdivided into following main categories:

- Crowbar protection scheme
- Fault Current Limiter scheme
- Series or parallel compensator scheme
- Energy storage protection scheme

3.1.1 Crowbar Protection Scheme

The crowbar protection scheme, shown in Fig. 2 for a DFIG-based wind turbine, allows the wind turbines to stay connected to the grid during grid faults limiting the high current in the rotor and providing a bypass for it via a set of resistors that are connected to the rotor windings. When the fault occurs in the grid, the current flows through short-circuit resistances rather than the rotor and the converter. The thyristors are connected to the resistors of the rotor circuit. Therefore, the rotor side converter (RSC) that is connected to the rotor of generator is not required to be disconnected from the DFIG when the grid faults happen. There are many advantages of this protection scheme including

- The RSC is not disconnected from the DFIG,
- The synchronism of operation remains during and after the fault, and normal operation can be continued immediately after the fault has cleared,
- The transients would not occur because the machine is not disconnected from the system,

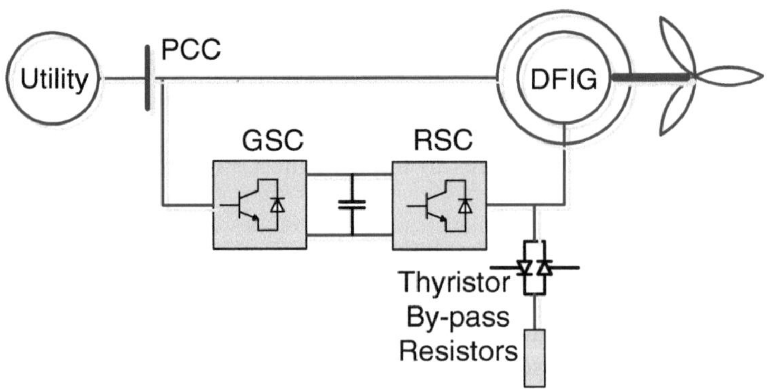

Fig. 2 DFIG bypass resistors in the rotor circuit

- Possibility to supply the reactive power to the system during the faults in order to support voltage restoration [5].

However, there are still some disadvantages of this protection scheme. For the period longer than few milliseconds, the crowbar needs to be disconnected from the system. It was suggested by the authors in [5] that this protection scheme would disconnect the rotor from the system if the voltage sag was greater than 0.2 pu grid voltage.

Protection of DC link voltage of a PMSG that is connected to the grid through a back to back converter is usually done by a DC chopper. It connects in parallel with DC capacitor and regulates its voltage by controlling its discharging into a resistor. Figure 3 shows a schematic diagram for a PMSG-based wind turbine that is equipped with DC chopper.

3.1.2 Fault Current Limiter Scheme

Fault current limiter (FCL) can be used to improve the LVRT of wind turbines, effectively. A FCL is a device which limits the prospective fault current when a fault occurs. Main characteristics of an ideal FCL are

- Zero impedance in the normal operation,
- No power loss in the normal operation,
- Large impedance in the fault conditions,
- Quick appearance of impedance when a fault occurs, and
- Fast recovery after fault removal.

Two main types of superconducting FCLs are used for LVRT improvement of wind turbines which are:

- Solid State switch-based FCL, and
- Bridge type FCL.

Figure 4 shows a sample application of solid state FCL (SSFCL) for LVRT improvement purpose. The SSFCL is in series with wind turbine. The SSFCL consists of a static switch (including four diodes and a gate-turn-off semiconductor

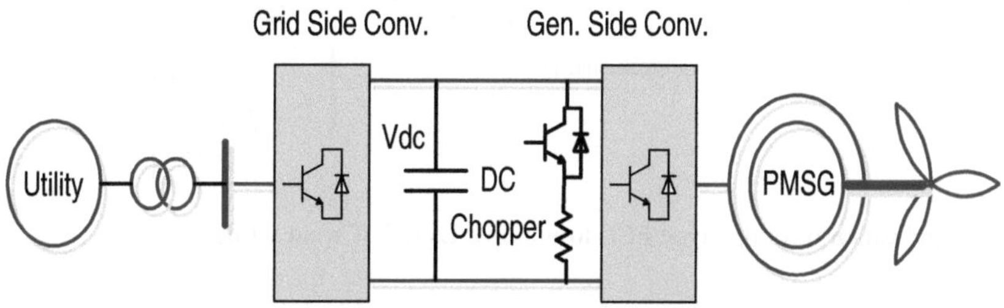

Fig. 3 PMSG-based wind turbine equipped with DC chopper

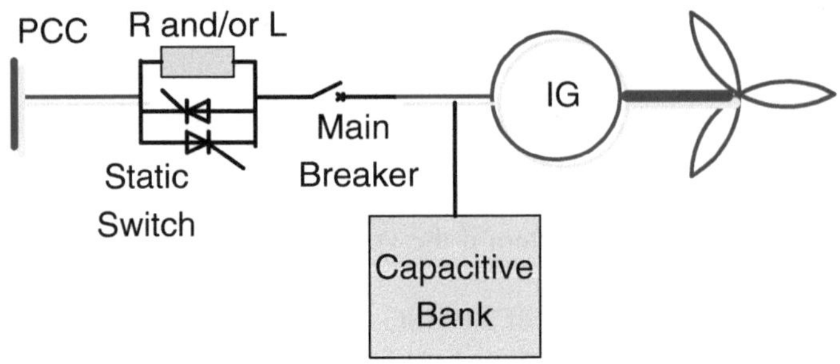

Fig. 4 The single machine infinite bus system with the induction generator-based WTPG [6]

switch) that is in parallel with impedance. The impedance might be a resistor or inductor for a resistive or inductive SSFCL types, respectively. In normal operation of utility, the static switch is ON and the parallel impedance is short circuited. In fault condition, the static switch is OFF and its parallel impedance connects in series with fault and limits the fault current. The resistive type FCL can be used to dissipate the power of wind turbine during fault and improve its stability characteristics.

Figure 5 shows a bridge type FCL that is connected in series in the feeder which connects the wind turbine to the grid. The bridge type FCL consists of four diodes and a DC inductor. During normal operation of utility, the current through DC inductor is almost DC and it becomes short circuited. During fault, the current starts to increase rapidly but the inductor limits the increase of fault current; and in this way, it decreases the voltage dip at point of common coupling (PCC) which improves the LVRT of wind turbine.

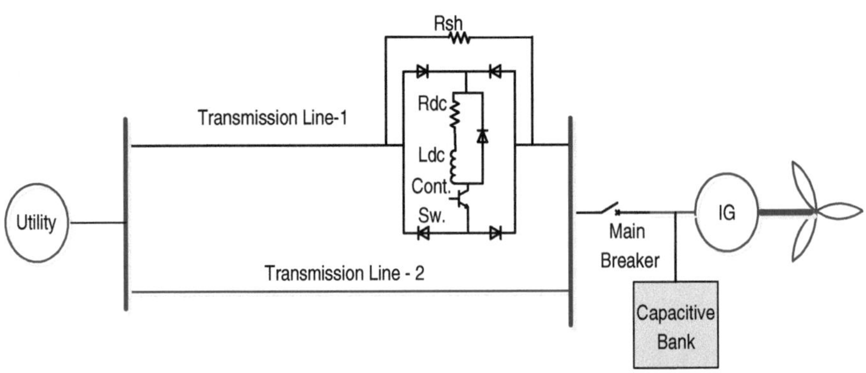

Fig. 5 Application of bridge type FCL to improve LVRT of wind turbine [7]

3.1.3 Series or Parallel Compensator Scheme

Voltage source inverters are used for improvement of LVRT of wind turbines as series or parallel compensators. These force commutated power electronic compensators connect to the utility either by an appropriate transformer or without transformer. The transformer-less configuration may use multilevel-type inverters or connect in parallel with utility through an appropriate inductor. The power circuit topology of series and parallel compensator schemes are explained below.

Fault condition results in a voltage sag in wind turbine terminal. Series compensators connect in series with wind turbine and inject a compensation series voltage for increasing of voltage value in wind turbine terminal during fault. The most common series power electronic compensator for this purpose is dynamic voltage restorer (DVR). Figure 6 shows a (DVR) that is connected in series with a DFIG-based wind turbine. The DVR injects a series voltage during fault to compensate for voltage sag at the terminal of the wind turbine point of common coupling (PCC). Compensation for voltage sag makes it possible for wind turbine to continue its active power injection to the utility during fault. This will prevent voltage rise of DC link (V_{dc}) due to excess of active power of wind turbine. It should be noticed that wind turbine should commit in providing fault current even with operation of DVR during fault condition.

Parallel compensators inject reactive power to the PCC of wind turbine and utility for voltage sag compensation during fault. The most common configuration is STATic COMpensator (STATCOM). Figure 7 shows a grid connected wind energy system with STATCOM. During fault, the STATCOM injects reactive power to the PCC and supports voltage. This condition improves FRT of wind turbine by providing the possibility for continuing active power injection to the utility during fault condition. The wind turbine as well as STATCOM should commit in providing fault current. The STATCOM could be used for reactive power absorbing during normal operation. It could be used as a parallel active filter for current harmonic compensation or dynamic stability improvement of wind turbine during normal operation of utility. Series or parallel compensator schemes could be combined with each other for the solution of LVRT [8, 9].

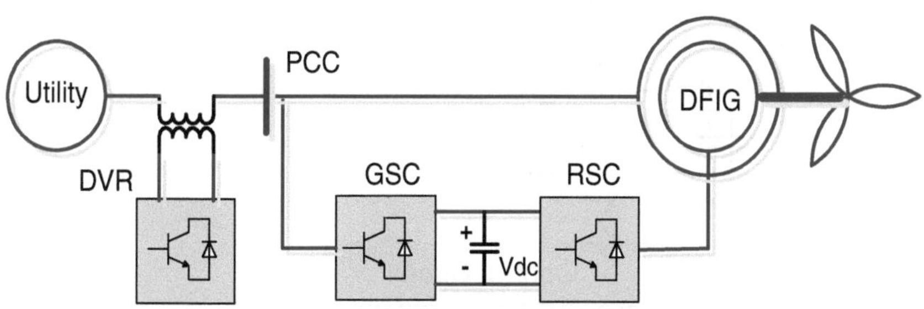

Fig. 6 Application of a DVR for LVRT improvement in DFIG-based wind turbine

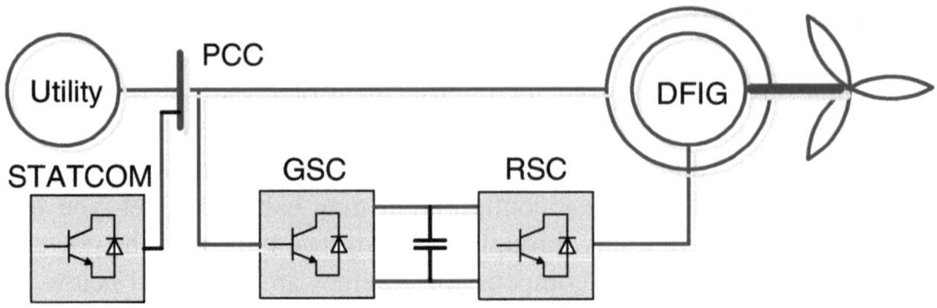

Fig. 7 Application of a STATCOM for LVRT improvement in DFIG-based wind turbine

3.1.4 Energy Storage Protection Scheme

Another solution to LVRT of wind turbine is an energy storage protection system. Figure 8 shows this strategy for a DFIG wind turbine. The energy storage system, connected to the DC link is providing an extra spinning reserve facility for the wind system, which can be used as a sink, when DFIG operates in super synchronous mode, or source, when DFIG operates in sub synchronous mode, of active power. The energy storage system can be used to smooth the stochastic wind variations so that the power output is smoothed in normal operation of utility which reduces voltage flicker too [10]. For simulation purpose, the wind turbine of Fig. 8 is connected through a 575/11 kV, 2 MVA step up transformer to 11 kV transmission line. The parameters used for wind turbine are as follows:

Wind turbine 1.5 MW, Power coefficient = 0.73, Speed = 1.2 pu

DFIG 1.5 MW, Nominal frequency = 50 Hz, Nominal voltage = 575 V, Stator resistance = 0.00706 pu, Stator inductance = 0.171 pu, Rotor resistance = 0.005

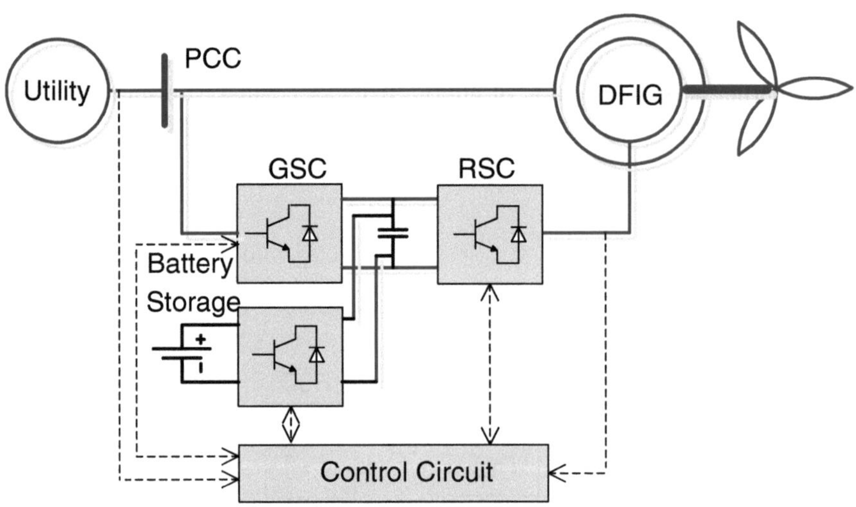

Fig. 8 Diagram of DFIG-based wind turbine with energy storage system [10] and [16]

pu, Rotor inductance = 0.156 pu, Magnetizing inductance = 2.9 pu, Inertia constant = 5.04 s, Friction factor = 0.01 pu, Number of pairs of poles = 3, Slip = 0.09 pu

RSC and Grid-Side Converter (GSC) Switching frequency = 1350 Hz, DC link voltage = 1200 V, Capacitor size = 10 mF.

A three-phase fault is considered at 18 km away from the PCC toward the utility. The control system of battery energy storage is designed to regulate the DC side voltage during fault. The LVRT capability is analyzed with energy storage such as 10 mF capacitor, 12 m/s wind speed and seven different voltage sags. From Fig. 9, it can be seen that the LVRT capability depends on the voltage sag and duration of that voltage sag. It can be seen that for higher voltage sags there is less time allowed for wind turbine to stay connected to the grid and vice versa.

Figure 10 shows the LVRT capability versus the capacity of energy storage and the duration for which the wind turbine is able to stay connected to the grid during the voltage sags before disconnection. For example, for energy storage of 5 mF capacitor, wind turbine can stay connected to the grid for 0.7 s. From this it can be concluded that with the inclusion of energy storage, wind turbine is able to stay connected to the grid during voltage sags and faults for a longer time. Also, energy storage can help reduce the DC link voltage, and real and reactive power fluctuations in the grid and DFIG machine.

Fig. 9 The LVRT capability for different voltage sags

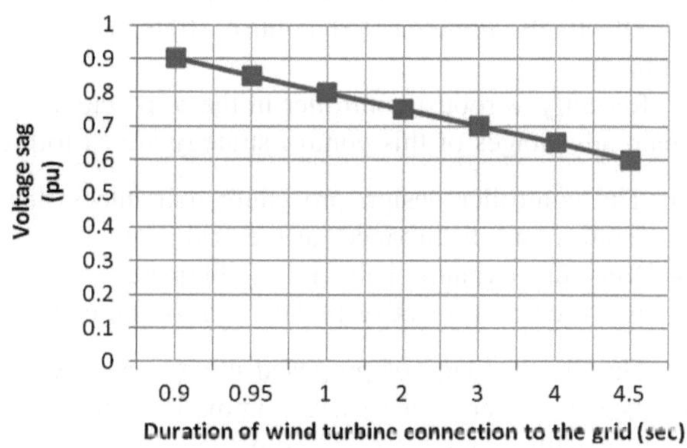

Fig. 10 The LVRT capabilities for different energy storages

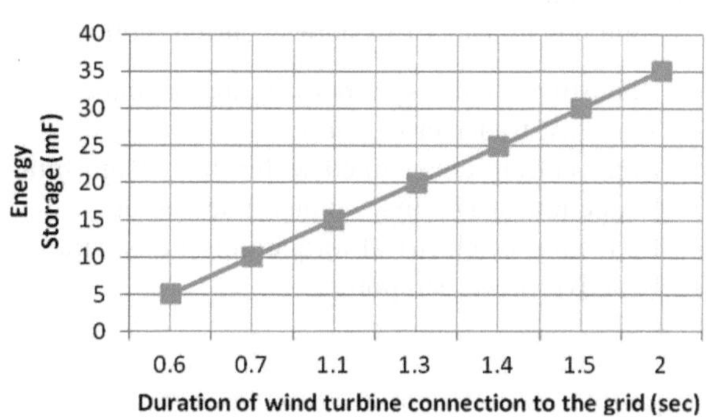

3.2 Control Strategy Based Solutions

Control strategy based solutions have acceptable LVRT improvement characteristics just for limited voltage dip values and duration; however, they are attractive because they do not need additional electrical equipment to be added to existing wind turbine power circuit topology. The cost for implementation of the control strategy may be cheaper than the cost for power circuit topology. Different control strategies are proposed in the literature while most of them are complicated and depend on the proper design of the control parameters [17–20]. These strategies are mainly based on using the following methods:

- Robust control,
- Hysteresis current controller,
- Artificial intelligent,
- Feedback linearization,
- Optimum design of PI controllers,
- Pitch control,
- Power angle control,
- Internal mode control, and
- Virtual crowbar.

Following parts of this section explains advantages and disadvantages of some of effectively implemented control strategies for LVRT improvement of wind turbines.

Recently, a robust controller in the α–β stationary frame is presented [17]. The main advantages of this control strategy are as follows:

- The controller design procedure guarantees the DFIG stability under uncertainties and disturbances at the grid side.
- Very fast dynamic behavior is obtained.
- Rotor current, reactive power, and torque control are achieved.

The disadvantage of proposed method is that the results are obtained with an oversized converter that can accommodate rotor overvoltage and full rotor current control with a suitable-sized converter, and hence this control method may have some limitations.

Main advantages of hysteresis current controller are [18, 19]:

- Robustness under symmetrical and asymmetrical fault conditions,
- Very fast transient response,
- Rotor current constrained within the design limits,
- Reduction of DC link voltage fluctuations,
- Reduction of electrical torque pulsations, and
- Reactive power support.

Hysteresis current controller has the following main disadvantages:

- Higher harmonic content,
- Higher switching frequency, and
- If the maximum switching frequency is limited, large error bands that produce significant low-order harmonics.

Computational intelligence-based control strategy for LVRT improvement has showed the following advantages [20]:

- Overcurrent at the rotor windings are effectively constrained below their maximum acceptable values.
- The DC link overvoltage is effectively constrained below their maximum acceptable values.
- In the case studied, the grid voltage, being supported by the proposed control system recovers quickly, so there is no need for large amounts of reactive power from the DFIG after the fault.
- The DFIG delivers the required amount of reactive power to the grid, in order to support the grid voltage.

Intelligence-based control strategies are in progress and some issues like their performance in asymmetrical faults and electrical torque pulsations could be investigated.

4 Synchronization Scheme for Wind Turbine Reconnection to the Grid

Disconnection of wind turbines after a fault refers to a major event in power system, especially if penetration of wind turbines which are disconnected from the utility during fault has been significant. During a major event, the disconnected wind turbines may not be in operation for an extended period of time. To minimize this effect, automatic synchronization should be implemented for reconnection or restoration of wind turbines as soon as possible after clearing the fault. This section deals with the challenges of resynchronization of wind turbines for their reconnection to the utility.

Synchronization is an extremely important aspect of all power systems. It refers to the requirement for the 3-phase power supplied by the generator to be exactly in phase with the 3-phase power supply on the grid. If synchronization is not achieved and the generator is reconnected to the grid, there is a high possibility that the two systems will be out of phase, which could lead to damage of components connected to the grid.

To be able to connect the generator to the grid it is required that the following three conditions are met [21]:

- Frequency is equal to system frequency.
- Voltage is equal to system voltage.
- Voltage is in phase with the system voltage.

In the islanded system, it is required that when the grid supply is restored, the wind generators will be allowed to automatically reconnect to the grid. A requirement of being able to do this is that the generator can automatically resynchronize with the grid supply before the connecting breaker is able to be closed.

Currently, there are various methods of resynchronization; these use the islanding control system, with additional control blocks to allow for synchronization between grid and stator voltage.

The control logic is based upon a normal technique of synchronization. Figure 11 shows the logic which has been used in the automatic synchronization technique. It can be seen from this figure that there are three main loops which are for synchronization of voltage magnitude, frequency, and voltage phase. Each of these loops compares the stator response with the grid response parameters. If their values are not the same, the control parameter is varied and the comparison is carried out again. This loop continues until the stator response is the same as the

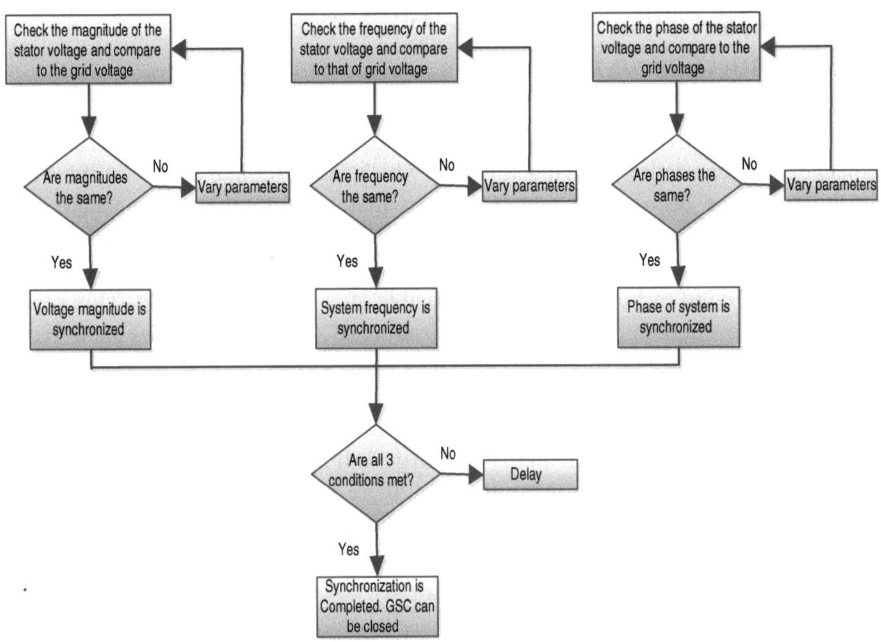

Fig. 11 Flowchart of synchronization logic

grid response. When these two values are equal it is known that the system is synchronized with the grid [21].

The final part of the control system checks that the three conditions are met and their corresponding signals are synchronized, if the three conditions are not met, a time delay will be triggered. At the end of this delay, the status of the three corresponding signals is checked again. This will continue until all three corresponding signals are synchronized. When the three conditions are met, the synchronization process is complete, and the wind turbine can be connected to the utility.

Power system of Fig. 12 is used for simulation of synchronization logic of Fig. 11. The simulation to be carried out involves the building of the islanded network with two DFIG-based wind farms. It is assumed that the load build up has been carried out, so that both wind farms are operating in islanding mode. It is required that synchronization between the two wind farms has to be carried out, that is, wind farms-1 acts as if it is the grid for wind farm-2. Once wind farm-2 is synchronized with the wind farm-1, the interconnection switch between the two systems, GSC-1 is closed. When this interconnection switch is closed, it is required that the wind farm-2 has to be put into the grid connected mode.

The final step is to synchronize the new island with the grid. This can be achieved by using the synchronization control mode to synchronize with grid voltage. Upon completion of synchronization with the grid, the grid connection switch, GSC-2, is closed and the both wind farm-1 and wind farm-2 mode is

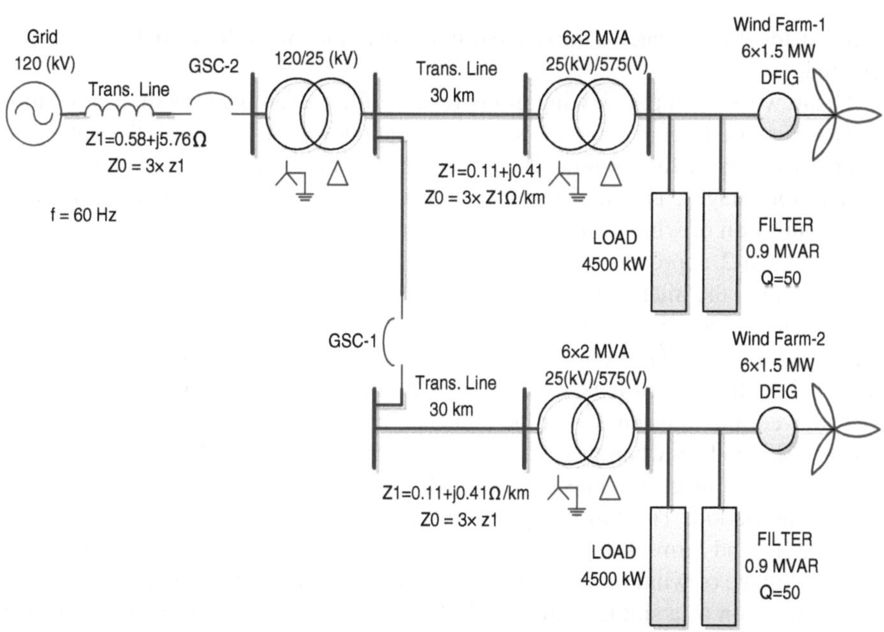

Fig. 12 Simulated power circuit to analyze resynchronization challenges of wind turbines

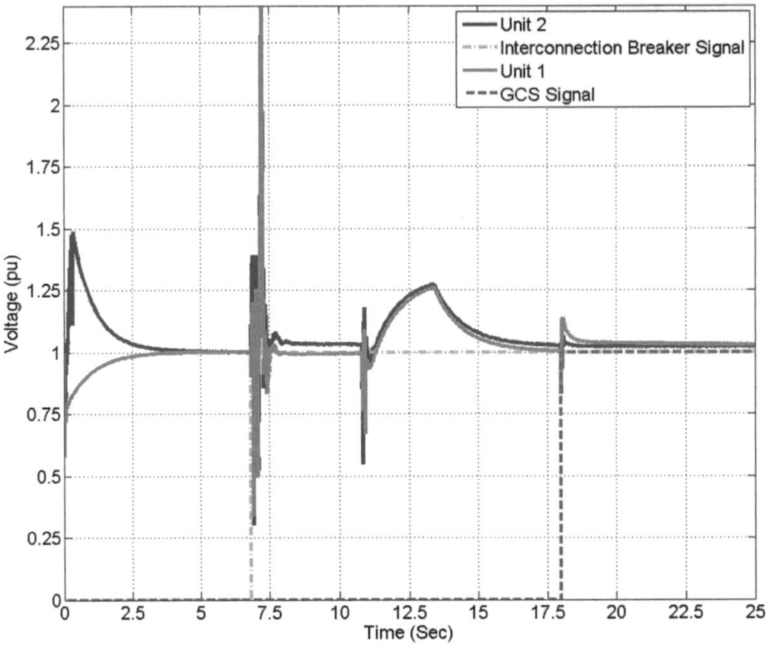

Fig. 13 Voltage over various stages of resynchronization of two islanded networks with each other and finally with the utility

changed to grid connected mode. Normal operation of both wind farms is now restored.

To allow for a simple viewing of the results of this example system, a simulation study was carried out to show the system operation. Figure 13 shows the voltage of several sections of the system. It can be seen in this figure that the voltage goes through four stages. The first stage occurs between 0 and 6 s, and in this period the synchronization of wind farm-2 and wind farm-1 occurs. It can be seen that voltage of wind farm-2 approaches the voltage of wind farm-1. Once they become very close to equal, the interconnection breaker GSC-1, is closed. This condition is used to simulate the possible delay of mechanical breakers to be closed exactly in same magnitude instant. This condition introduces a large voltage transient into the system. For this reason it is assumed that if complete process of synchronization was allowed, the transient would be reduced. The second stage of operation occurs in the time 7.5–11 s when the two wind farms operate as a single island, with wind farm-2 running in grid connected mode. It can be noted that the system is stable during this period. The third stage occurs in the period 11–18 s during synchronization of wind farms 1 and 2 with the grid voltage. From the graph it can be seen that the voltage of wind farm-1 appears to take the same course as was shown in the synchronization of a single system with the grid. For this, the transients introduced from closing of the GSC-2 and changing to grid connected mode are minimal. The fourth or final stage shown in Fig. 13 represents the period after the system has been reconnected to the grid. It can be seen that during this time normal operation of the system has been restored, which confirms that building of the islanded network is successful.

5 Conclusion

This chapter focuses on the importance of low voltage ride through (LVRT) of wind turbines. The investigation of some selected grid codes has revealed that LVRT has different requirements in different grid codes. The results of the investigation have found that higher penetration of wind energy in grid results in stricter LVRT requirements. On the other hand, increasing the wind energy penetration in grid is an ongoing process, so it will be mandatory to increase the LVRT performance of existing wind turbines as well as newly installed units. This chapter has presented some key power circuit topology based and control strategy based LVRT improvement methods. While power circuit topology based strategies are usually more efficient than control strategy based methods, they need extra power electronic-based electrical circuits which might be expensive or bulky and they may affect the normal operation of wind turbines, too. This chapter has also discussed the problems of resynchronization of wind turbines. It is shown that without sufficient LVRT capability, wind turbines would be disconnected from grid during faults and resynchronization of them would be a time-consuming process which may impose large transients to the grid. A resynchronization scheme for automatic reconnection of wind turbines has been presented and validated through simulation.

References

1. http://en.wikipedia.org/wiki/Low_voltage_ride_through
2. "ENTSO-E Network Code for Requirements for Grid Connection Applicable to all Generators", 8 March 2013
3. Australian government, Bureau of resources and energy economics, July 2014
4. "National Electricity Rules", Version 63, Chapter 5, Network Connection, Planning And Expansion, published on 1 July 2014, pp 371–593
5. Morren J, de Haan SWH (2005) Ride through of wind turbines with doubly-fed induction generator during a voltage Dip. In: IEEE Trans Energy Convers 20(2):435–438
6. Fereidouni R, Vahidi B, Hosseini Mehr T (2013) The impact of solid state fault current limiter on power network with wind-turbine power generation. IEEE Trans Smart Grid 4(2)
7. Rashid G, Hasan Ali M (2014) A modified bridge-type fault current limiter for fault ride-through capacity enhancement of fixed speed wind generator. IEEE Trans Energy Convers 29(2):527–534
8. Huang P, El Moursi MS, Weidong X, Kirtley JL (2013) Novel fault ride-through configuration and transient management scheme for doubly fed induction generator. IEEE Trans Energy Convers 28(1):86–94

9. Alaraifi S, Moawwad A, El Moursi MS, Khadkikar V (2013) Voltage booster schemes for fault ride-through enhancement of variable speed wind turbines. IEEE Trans Sustain Energy 4 (4):1071–1081
10. Aktarujjaman M, Kashem MA, Ledwich G (2016) Smoothing output power of a doubly fed wind turbine with an energy storage system. In: IEEE Power electronics, drives and energy systems, 2006. PEDES '06. International conference, 12–15 Dec 2006, pp 2–6
11. Li H, Chen Z (2006) A new current control strategy of maximizing the generated power from a doubly fed induction generator system. In: 12th international power electronics and motion control conference, IEEE, EPE-PEMC, Portoroz, Slovenia, Aug 2006, pp 1557
12. Pors A (2006) APSCAD™/EMTDC® Model for the doubly-fed induction generator wind turbine scheme. School of Electrical, Computer and Telecommunications Engineering, University of Wollongong, ECTE457, Thesis, Oct 2006
13. Muller S, Deicke M, De Doncker RW (2000) Adjustable speed generators for wind turbines based on doubly-fed induction machines and 4-Quadrant IGBT converters linked to the rotor. In: Conference record of the 2000 IEEE industry applications conference 2000, vol 4, Oct 2000, pp 2250–2253
14. Mohan, Underland, Robbins (2013) Power electronics converters. applications, and design, 3rd edn, Chap 6. John Wiley and Sons
15. Lei Y, Mullane A, Lightbody G, Yacamini R (2006) Modeling of the wind turbine with a DFIG for grid integration studies. IEEE Trans Energy Convers 21(1):258
16. Aktarujjaman M, Kashem MA, Negnevitsky M (2006) Improvement of output of doubly fed wind turbine via storage system. In: IEEE International Power and Energy Conference, 2006. PECon '06, pp 312–315, 28–29 Nov 2006
17. da Costa JP, Pinheiro H, Degner T, Arnold G (2011) Robust controller for DFIGs of grid-connected wind turbines. IEEE Trans Ind Electron 58(9):4023–4038
18. Mohseni M, Islam S, Masoum MAS (2011) Fault ride-through capability enhancement of doubly-fed induction wind generators. IET Renew, Power Gen 5(5):368–376
19. Mohseni M, Islam S (2011) Transient control of DFIG-based wind power plants in compliance with the Australian grid code. IEEE Trans Power Electron 27(6):2813–2824
20. Kanjiya P, Ambati BB, Khadkikar V (2014) A novel fault-tolerant DFIG-Based wind energy conversion system for seamless operation during grid faults. IEEE Trans Power Syst 29(3)
21. Wildi T (2006) Electrical machines, drives and power systems, 6th edn. Pearson Education, New Jersey

Chapter 4
Reliability Evaluation of Distribution System with Network Reconfiguration and Distributed Generations

P. Pavani and S.N. Singh

Abstract This chapter titled "Reliability Evaluation of Distribution System with Network Reconfiguration and Distributed Generations" proposes reliability and power loss evaluation of distribution system using network reconfiguration tool and optimal DG siting and sizing. The objectives are reliability improvement and power loss reduction subjected to various operating constraints. The renewable DG technologies including the synchronous machine-based biomass generator are considered in distribution system expansion planning. The network reconfiguration with DGs and tie-switch placement for the reliability improvement and loss minimization is proposed. Tie-switches are placed at terminal nodes with geographical constraints and their all possible binary combinations are considered. DGs sizing for the predefined DGs are calculated using integer programming for objective viz. reliability improvement and minimization of losses with penetration level taken for less than total load. Power loss is calculated using forward/backward sweep distribution load flow algorithm in all the cases analyzed. A search-based reconfiguration algorithm has been formulated for finding the optimal switch configuration for the radial distribution system with DG. The optimal DG placement and DG sizing for combined objectives of reliability improvement and loss minimization, using fuzzy logic approach is proposed in two steps. Out of four DG variables, type, and number are taken as fixed variables and size and location are considered as variables in the formulation. In the first step, for all DGs combination, their optimal sizes are calculated using nonlinear optimization tool with the improved reliability and real power loss reduction objectives. SAIDI and SAIFI indices are used for the reliability evaluation in this case. In the formulation three indices, power loss, SAIDI and SAIFI are used with weighing factors and taken as combined objective. A combined network reconfiguration and optimal DG siting and sizing is carried out for the objectives viz. real power loss reduction and reliability improvement for various penetration levels and various loading conditions. It is observed that optimal sitting and sizing of the DGs improve the system performance in terms of real

P. Pavani · S.N. Singh (✉)
Department of Electrical Engineering, Indian Institute of Technology Kanpur, Kanpur, India
e-mail: snsingh@iitk.ac.in

N.R. Karki et al. (eds.), *Sustainable Power Systems*, Reliable and Sustainable Electric Power and Energy Systems Management,
DOI 10.1007/978-981-10-2230-2_4

power loss reduction and improvement in the system reliability. Different objectives give different optimal locations and sizes but based on the physical constraints, the final decisions are to be made. The present work may be a guideline to the planners and policy makers for planning the DGs in the distribution systems.

1 Introduction

Modern power systems are undergoing major changes in terms of ownership of assets, operation, and management. The current era is focusing to make and operate the electric grid “smart” in all possible aspects. The main objective of the power systems is to transmit power reliably, securely, efficiently, and economically to the end users. Distribution automation allows flexible control of distribution systems to the utilities, which can enhance the efficiency, reliability, and quality of electric service. One important application of distribution automation can be distribution network reconfiguration. Network topology changes by the opening or closing of the sectionalizing or tie-switches constitutes network reconfiguration, while meeting the load requirements and radial nature of the network [1]. Due to tremendous growth in power demand, rising environmental concerns and requirement of extensive transmission and distribution networks, the electric utilities have chosen the distributed generators (DGs) at the load end. DGs gain advantage over the conventional grid power because of their small size and closeness to the end users thus avoiding the transmission and distribution costs. DGs are also often an unbeatable alternative for electric power needs in some remote and sparsely populated situations. In rural areas, where electric service points are far off, even a power distribution system of the modest capacity will cost much. By contrast, DGs placed at just the selected sites can give economic solution of less than half the total cost. Economic factor is one of the most important elements in nearly all distributed generation evaluations, although alone, it is insufficient for complete evaluation. Both reliability and power quality are essential elements of the decision [2]. DGs, especially renewable energy-based distribution generations are widely installed in the distribution systems due to various economic reasons. The increasing penetration of DGs also improves system reliability and service quality. Installation of DGs at or near end users can benefit electric utilities by avoiding or reducing transmission and distribution system upgrades. So appropriate siting and sizing of DGs can reduce power losses and can improve the system reliability.

It is well known that the voltage profiles along a loaded distribution network feeder are typically such that the voltage is high close to the distribution transformer bus-bar, and the voltage drops along the length of the feeder due to radial nature. The placement of DGs in a distribution feeder usually reduces the voltage drop along the feeder, and may lead to voltage rise, if reactive power produced by DG, at some points. The level of voltage rise depends on generation penetration level compared to the minimum load demand. In [3], genetic algorithm (GA) is used for DG allocation for losses, reliability, and voltage improvement. Particle swarm optimization (PSO)-based multi-objective algorithm for DG siting is proposed in [4]. A combined GA and PSO-based algorithm is proposed in [5] for placement and sizing of DG considering several objectives such as power loss reduction, voltage profile improvement, and maximizing voltage stability. A continuation power flow-based DG placement algorithm for voltage stability margin improvement at various DG penetration levels has been developed in [6]. With DG integration, distribution system losses can be reduced with better reliability. Network reconfiguration also aims at reducing losses and improving system reliability [7]. The switch placement and network reconfiguration with DG are carried out in [8] for improving the system reliability. The owner/operator of a DG system can specify some amounts of redundancy, to create higher levels of power availability (reliability) and power quality (e.g., voltage regulation) than that provided by the electric system. These are done on a site-specific basis. DG placement for the reliability improvement has been proposed in [9]. In [10], network reconfiguration for power loss reduction using harmony search algorithm is proposed and simultaneous DG placement and network reconfiguration for power loss reduction is proposed in [11]. The complete literature review on DG placement that had been carried out till now is presented in [12]. DG placement is carried out considering uncertainties using GA and Monte Carlo simulations in [13, 14]. In [12, 13], the DG placement is carried out using evolutionary algorithms and presented the convergence characteristics of the algorithm.

The DGs are integrated into the utility grid as per IEEE-1547 standard [15]. The fault levels, performance of protection devices and power quality can adversely be affected by large penetration of the DG [16]. The DGs must be placed strategically in the system to reduce the network losses, improve system reliability, and efficiency [17]. The DG location, size, and type are the three main factors that should be selected appropriately otherwise it may affect the system performance adversely [18]. In [19], modal analysis and continuation power flow-based DG placement is proposed for the voltage stability, however, stability problem is very rare in the radial distributed systems. A multi-objective nonlinear programming (NLP)-based method is proposed in [20, 21] to find optimal size and to locate DGs in both meshed and radial power systems, for minimizing power losses as well as maximizing the voltage stability margin

In [22], a method using the Kalman filter algorithm is proposed to determine the optimal locations and sizes of multiple DGs in a practical distribution system. A two-part methodology was developed in [23] that ranks utility feeders for the best candidates for distributed energy resources (DER) and then optimal placement and sizing of DER based on technical and economic considerations are carried out. Siting and sizing of DG to maximize penetration level from both synchronous type and inverter-type of DG units, taking power balance constraints, bus voltage limits, and harmonic distortion limits is proposed in [24]. An optimization methodology for optimal sizing and sitting is developed in [25], to enhance the loadability of the distribution system by improving static voltage stability. A hybrid PSO-based DG placement is proposed considering voltage rise problem in [26]. The investment and operation of DG units on power loss under uncertainties is evaluated using fuzzy tool in [27].

This work proposes reliability and power loss evaluation of distribution system using network reconfiguration tool and optimal DG siting and sizing. The objectives are reliability improvement and power loss reduction subjected to various operating constraints, which are carried out using three different ways described as follows:

- Network reconfiguration with DG
  - Tie-switches are placed and DGs are already present in the system.
  - Reconfiguration algorithm is used for reliability improvement and loss reduction.
- Optimal DG siting and sizing
  - Out of four DG variables, i.e., type, location, number and size, type and number are fixed and rest two are found using optimization tool.
  - A fuzzy min-max approach is utilized for finding the optimal DG siting and sizing for the reliability improvement and loss reduction.
- Network reconfiguration and optimal DG siting and sizing.
  - Reconfiguration and optimal DG siting and sizing is carried out simultaneously.
  - Analytical hierarchy process is utilized for finding the optimal DG siting and sizing for any system configuration and loadings.

It is assumed that all the test systems are fully automatized with sectionalizing switches in all the lines. The effectiveness of the proposed approach is tested on 33-bus and 69-bus distribution systems.

2 Reliability Indices

Electric utilities are expected to provide continuous and quality electricity at a reasonable rate to the end users by doing economical justice to the available resources. Reliability describes the security of the system and the service continuity between supply and load. The classical definition [1] of reliability is the probability of a device or system performing its function adequately, for the period of time intended, under the given operating conditions. The path from load to supply is termed as minimal cutset and the unavailability of any component in this path will correspond to the loss of load [2]. Until unless there is an outage of the components connecting load to supply, there exists a path between load and supply. To measure system performance, the electric utility has developed several performance measures of reliability. These reliability indices include measures of outage duration, frequency outages, system availability, and response time. The basic single component concept is as follows.

$$\mathrm{MTTF} = \sum_i \frac{(1/\lambda_i)}{n} \quad (1)$$

$$\mathrm{MTTR} = \sum_i \frac{(1/u_i)}{n} \quad (2)$$

The availability of a component (P) is given by

$$P = \frac{\mathrm{MTTF}}{\mathrm{MTTF} + \mathrm{MTTR}} \quad (3)$$

The unavailability (Q) of the component can be written from Eq. (4) as

$$Q = 1 - P \quad (4)$$

The overall system average unavailability (Q_{as}) given in Eq. (5) can be calculated where L is total number of loads and Q_i unreliability at load point and the basic assumption in using the Eqs. (1)–(5) is that the probability of availability exponentially decreases with time.

$$Q_{\mathrm{as}} = \frac{1}{L}\sum_{i=1}^{L} Q_i \quad (5)$$

MTTF and MTTR are the mean time to failure and mean time to repair of ith component, respectively. λ, u and n are the failure rate, repair rate, and number of cycles of the component. Distribution system contains large number of components connected either in series or parallel. The minimal cutsets can be of single order, second order, and so on. If a load is connected to the mains by two components connected in series, the unavailability of any one component will make the load disconnected from supply. If the components are in parallel, then the probability of loss of supply to load will be unavailability of both components, i.e., if any one component is available then continuous supply to the load can be maintained. The reliability can be improved by the DGs due to continuity of supply to loads even under contingencies. Interruptions and outages can be studied through the use of predictive reliability assessment tools that can predict customer reliability characteristics based on system topology and component reliability data. In order to achieve this, distribution reliability indices including both duration and frequency of outage are calculated. Network reliability assessment contains two types of indices: system indices and load point indices. Load point indices are calculated at each load point connection and are used for the evaluation of system indices. The load point indices are energy not supplied (ENS) and average energy not supplied

(AENS). The system indices that can be considered for reliability evaluation are system average interruption frequency index (SAIFI) and system average interruption duration index (SAIDI) defined as.

$$\text{SAIDI} = \frac{\sum u_i \times N_i}{\sum N_i}, \tag{6}$$

$$\text{SAIFI} = \frac{\sum \lambda_i \times N_i}{\sum N_i}, \tag{7}$$

where N_i is the number of customers at the load point i. The optimization algorithm is used to minimize the weighted composite index including interruption duration and frequency components. The smaller the value of the defined reliability index (i.e., objective function), the higher is the system reliability.

3 Effect on Power Loss

The integration of DG into the system considerably reduces the power losses in the system. It can help the utilities to meet the peak load requirements and can also avoid the unwanted load shedding. With proper placement and size of DGs, the system voltage profile and reliability can also be enhanced [19]. Under system intact conditions, network configuration tool is helpful to find the possible lowest loss path from load to supply. This can also be useful during contingency by configuring the distribution system in such a way that least number of loads will undergo outage which can improve reliability of the system. Forward/backward sweep distribution load flow algorithm is used and DG is modeled as PQ nodes, i.e., negative loads. The loss function (P_{loss}) is given as

$$P_{\text{loss}} = \sum_{j=1}^{l} r_j \frac{P_j^2 + Q_j^2}{|V_j|^2}, \tag{8}$$

where r_j is the resistance of the line j, V_j is the voltage at the receiving end of the line j, P, Q are total real and reactive power loads fed from receiving node of the line j.

4 Proposed Methodology

It is assumed that all the test systems are fully automatized with sectionalizing switches in all the lines. All the components are assumed to have identical reliability data with failure rate, repair time, maintenance rate, and maintenance time as 0.02 outages/year, 30 h, 0.2 outages/year, and 20 h, respectively. The upper and lower voltage limits considered are 1.0 and 0.95 p.u., respectively. The objectives are reliability improvement and power loss reduction subjected to various operating constraints, which are carried out using three different cases described as follows:

- Network reconfiguration with DG [28]
- Optimal DG siting and sizing
- Network reconfiguration and optimal DG siting and sizing, simultaneously

4.1 Case 1

In this case, network reconfiguration is used for reliability and power loss evaluation with DGs present in the network. Initially, tie-switches are placed at end nodes considering geographical constraints. The data for the tie-switches are considered same as the lines connected adjacent to them. The proposed search algorithm searches for the optimal network reconfiguration of the radial distribution systems with distributed generation to improve reliability and reduce losses. The objectives are the minimization of real power losses and improvement of reliability subjected to the voltage limits, power balance equations, and the real and reactive power generation of DG must be greater than losses and less than total system load as given in Eqs. (9)–(13). The variables in this formulation are DGs size and calculated values are the objectives.

Minimize

$$(P_{\text{loss}} + Q_{\text{as}}) \tag{9}$$

$$\sum_{j=1}^{n} P_{\text{DG}j} \leq C, \tag{10}$$

$$P_i = P_{Gi} - P_{Di}, \tag{11}$$

$$Q_i = Q_{Gi} - Q_{Di}, \tag{12}$$

$$0.95 \leq V_j \leq 1.05, \tag{13}$$

where P_{DG_i} is the real power capacity of *j*th DG, P_i, P_{Gi}, P_{Di} and Q_i, Q_{Gi}, Q_{Di} are the real and reactive, power injection, generation and demand at *i*th node, respectively and C is the total load in the system. The final step is the network reconfiguration considering reliability and power loss. In this step, all the loops formed by closing the tie-switches and the minimal cutsets are found. Constraints for this problem are maintaining system radial nature, power balance equations, and voltage limits. Distribution load flow algorithm is used for the power loss calculation and reliability evaluation models are used for calculating system reliability, downtime. The proposed methodology is explained using the following example shown in Fig. 1.

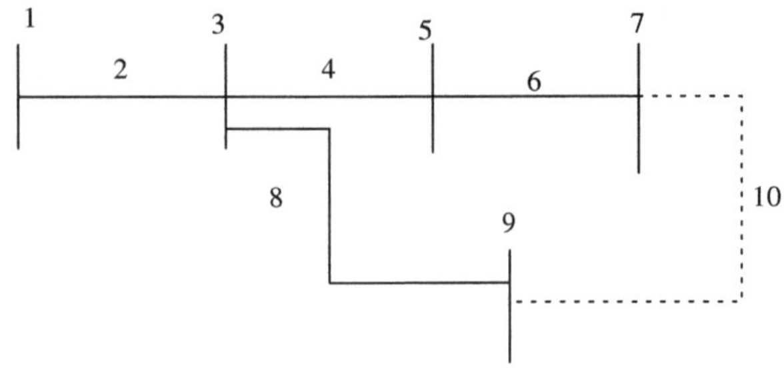

Fig. 1 Example distribution system

The terminal nodes for the example system are nodes 7 and 9. Tie-switch is placed between the nodes 7 and 9, which will be in open position. DG is present at node 7. The optimal DG size obtained is, $P_{\mathrm{DG}} = 103$ kW satisfying all the constraints. Now search-based reconfiguration problem will be carried considering both reliability improvement and power loss minimization which is multi-objective in nature. Initial step is collecting the required information, i.e., reliability data, tie-switches placed. The minimal cutsets are found using dynamic programming with and without considering tie-switches. Using probabilistic reliability evaluation method, the unavailability of all the load nodes without tie-switch 7–9 are given in Table 1. The loop formed by closing tie-switch 7–9 is 3–4–5–6–7–10–9–8 and is given in Table 2. With reconfiguration algorithm, the optimal configuration is opening 5–7 and closing 7–10 switches, respectively. The probability of average system unavailability came out to be 0.05616 %.

4.2 Case 2

In this case, evaluations of reliability and power loss are carried out using optimal DG placement and sizing. The DG penetration level is considered as 35 %. The DG number is fixed to four and type is fixed to synchronous generator type DG. All the possible combinations of DG as a set of four are considered and the problem formulation for the optimal DG sizing for all possible DGs configurations is given in (14)–(20) as,

Table 1 Unavailability of load nodes

Output node	Load to supply paths	Unavailability (%)
3	3, 2, 1	0.03120
5	5, 4, 3, 2, 1	0.05200
7	7, 6, 5, 4, 3, 2, 1	0.07280
8	9, 8, 3, 2, 1	0.05200

Table 2 Loops formed by closing tie-switch

Loops	Lines involved in loop
3–4–5–6–7–10–9–8–3	3–5, 5–7, 3–9, 7–9

$$\text{Minimize } (w_1\text{SAIDI} + w_2\text{SAIFI} + w_3P_{\text{loss}}), \tag{14}$$

$$\text{subject to } P_i = P_{Gi} - P_{Di} = \sum_{q=1}^{n} V_iV_qY_{iq}\cos(\delta_i - \delta_q - \theta_{iq}) \tag{15}$$

$$Q_i = Q_{Gi} - Q_{Di} = \sum_{q=1}^{n} V_iV_qY_{iq}\sin(\delta_i - \delta_q - \theta_{iq}), \tag{16}$$

$$\sum_{i=1}^{n} P_{\text{DG}_i} \leq 0.35 \times P_{\text{main}} \tag{17}$$

$$P_{\text{DG}_i}^{\text{min}} \leq P_{\text{DG}_i} \leq P_{\text{DG}_i}^{\text{max}}, \tag{18}$$

$$S_l \leq S_l^{\text{lim}}, \tag{19}$$

$$\sum_{k=1}^{n} w_k = 1\; w_k \in [0, 1], \tag{20}$$

where P_{main} is real power from the main substation, $P_{\text{DG}_i}^{\text{min}}, P_{\text{DG}_i}^{\text{max}}$ are minimum and maximum limitation on real power generation of DG at ith node, θ_{iq} is angle of the $i-q$th element of the bus admittance matrix, n is total buses in the system, δ_i is phase angle of complex voltage at node i, Y_{iq} is the $i-q$th element of the bus admittance matrix, S_l, S_l^{lim} are real power flow and thermal limit in the feeder, respectively. Three cases are considered in which Cases 1–2 represents single objective problems and Case 3 represents the multi-objective problem, in which the weights are as follows:

Case a: $w_1 + w_2 = 1\,\&, w_3 = 0$

Case b: $w_3 = 1\,\&, w_1 = w_2 = 0$

Case c: $w_1 + w_2 + w_3 = 1; w_1 + w_2 = 0.5,\ w_3 = 0.5$

In case c, reliability and power loss both were given equal weightage (0.5). For all the DG configurations, the optimal sizing is found. For each predefined DG configuration and its corresponding optimal size, the obtained power flow and reliability objective values are given to fuzzy logic min-max approach for finding the optimal DGs placement. The membership functions μPL_i, μR_i that correspond to the power loss reduction, and improved reliability, respectively, are calculated for all DG configurations.

4.2.1 Fuzzy Logic Approach

The process of formulating the mapping from a given input to an output using fuzzy logic is the fuzzy inference. The mapping then provides a basis from which

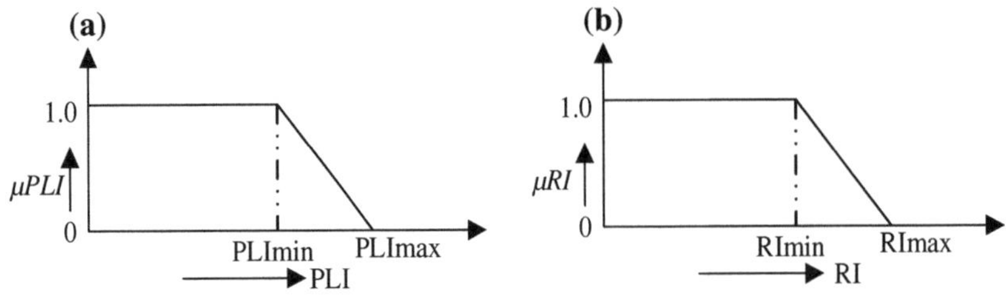

Fig. 2 Membership functions for **a** maximum power loss reduction, **b** reliability improvement

decisions can be made. The process of fuzzy inference involves: membership functions, logical operations, and if-then rules. A membership function (MF) defines how each point in the input space is mapped to a membership value (or degree of membership) between 0 and 1. A fuzzy set is a set without a crisp, clearly defined boundary. It can contain elements with only a partial degree of membership. The membership functions of the objectives are described below. This paper uses the fuzzy framework for the multi-objective problem based on min-max problem [26]. The membership functions are shown in Fig. 2.

Real Power Loss Membership Function

The real power loss index (PLI) is given as

$$\mathrm{PLI}_i = \frac{P_{\mathrm{loss}}^{\mathrm{base}} - P_{\mathrm{loss}}^{\mathrm{DG}_i}}{P_{\mathrm{loss}}^{\mathrm{base}}}, \tag{21}$$

where $P_{\mathrm{loss}}^{\mathrm{base}}, P_{\mathrm{loss}}^{\mathrm{with\,DG}_i}$ are power loss in the system without DG and with ith DG combination. The purpose of this membership function is to evaluate the real power loss reduction in the network. If PLI is high, the loss reduction is high, and high membership function is assigned, and if PLI is low, the loss reduction is low. The membership function for the real power loss (μPL_i) reduction is given below:

$$\begin{aligned} \mu\mathrm{PL}_i &= \frac{\mathrm{PLI}_{\max} - \mathrm{PLI}_i}{\mathrm{PLI}_{\max} - \mathrm{PLI}_{\min}}, \quad \text{for } \mathrm{PLI}_{\min} \le \mathrm{PLI}_i \le \mathrm{PLI}_{\max} \\ &= 1, \ \text{for } \mathrm{PLI}_i \le \mathrm{PLI}_{\min} \\ &= 0, \ \text{for } \mathrm{PLI}_i \ge \mathrm{PLI}_{\max} \end{aligned} \tag{22}$$

Reliability Membership Function

The reliability index (RI) is given in (23). The smaller value of the reliability index shows the greater the system reliability.

$$\mathrm{RI}_i = \frac{\mathrm{SAIDI_{base}} - \mathrm{SAIDI_{with\,DG_i}}}{\mathrm{SAIDI_{base}}} + \frac{\mathrm{SAIFI_{base}} - \mathrm{SAIFI_{with\,DG_i}}}{\mathrm{SAIFI_{base}}}, \tag{23}$$

where $\mathrm{SAIDI_{base}}, \mathrm{SAIDI_{with\,DG}}$ are SAIDI without DG and with DG, respectively, $\mathrm{SAIFI_{base}}, \mathrm{SAIFI_{with\,DG}}$ are SAIFI without DG and with DG, respectively. The purpose of this membership function is to see the reliability improvement. The weights for both the indices are assumed to be equal. If RI is high, the reliability is high, and a high membership function is assigned. If RI is low, reliability is low. The membership function for the reliability is given below:

$$\begin{aligned} \mu R_i &= \frac{\mathrm{RI_{max}} - \mathrm{RI}_i}{\mathrm{RI_{max}} - \mathrm{RI_{min}}}, \quad \text{for } \mathrm{RI_{min}} \le \mathrm{RI}_i \le \mathrm{RI_{max}} \\ &= 1, \ \text{for } \mathrm{RI}_i \le \mathrm{RI_{min}} \\ &= 0, \ \text{for } \mathrm{RI}_i \ge \mathrm{RI_{max}} \end{aligned} \tag{24}$$

In fuzzy set theory [27], the logical 'and' operator corresponds to the intersection of sets which is used in this work. The overall degree (OD) of satisfaction is the minimum (and) of all the above membership values.

$$\mathrm{OD}_i = \min\{\mu \mathrm{PL}_i, \mu R_i\} \tag{25}$$

The optimal solution (OS) is the maximum (or) of all such overall degrees of satisfaction.

$$\mathrm{OS}_k = \max\{\mathrm{OD}_p\} \tag{26}$$

4.3 Case 3

The procedure followed for finding the best DG locations and network reconfiguration, considering reliability improvement and loss reduction as objective, is illustrated in the below steps.

- Tie-switches are introduced in the system at the terminating nodes with geographical constraints taken into consideration. Different open/close binary combinations of the tie-switch set are generated.
- Various combinations of DGs are generated as a set of four. For each set of DG combination, optimal DG sizing is carried out for power loss minimization and reliability improvement satisfying the power balance equations and voltage limits, maximum, and minimum DG size limitations as the main constraints.
- A search-based reconfiguration algorithm is proposed and is applied on each combination of tie-switch set and DGs set for obtaining the best configuration of switches and optimal DG locations.

- Step 3 is repeated for various DG penetration levels, 20, 30, and 40 % of total system load.
- Step 4 is repeated for various loadings, i.e., 20 % of both load increase and decrease.
- AHP tool is used for obtaining the best configuration out of all tie-switch combinations, and the best DG location under different penetration levels and various system loadings for any system configuration considering improving reliability and reducing loss as the main objectives.

4.3.1 AHP Review

AHP allows for the application of the data, experience, insight, and intuition in a logical and thorough manner. The methodology enables the decision-makers to derive ratio scale of priorities or weights, as opposed to arbitrary assignments of weights, which may be assigned to the criteria. The decision making in AHP is a continuous process starting from analyzing the decision to understanding and then arranging the criteria into different groups and levels, and then finally evaluating the criteria in its decision outputs. AHP methodology is composed of several previously existing concepts such as hierarchy structuring of complexity, pairwise comparisons, redundant judgments, eigenvector method or deriving weights, consistency considerations, etc. The performance sensitivity (PS) or global priority (pi) is determined by synthesizing the local priorities across all criteria. The historical AHP approach (called later distributive mode) adopts an additive aggregation given in Eq. (27) with normalization of the sum of the local priorities (l_{ij}) along with their weights (w_j) to unity [29].

$$p_i = \sum_j w_j \cdot l_{ij} \tag{27}$$

5 Simulation Studies

5.1 33-Bus Distribution System

The simulation results are demonstrated on IEEE 33-node test distribution systems and problem is solved using *Fmincon* nonlinear optimization solver in Matlab. The single-line diagram of the IEEE 33 node test system is shown in Fig. 3 [30]. The total real and reactive power load on the system is 3715 kW and 2300 kVar, respectively. At base case condition, the maximum and the minimum bus voltage magnitudes are 1.0 and 0.898 p.u., respectively.

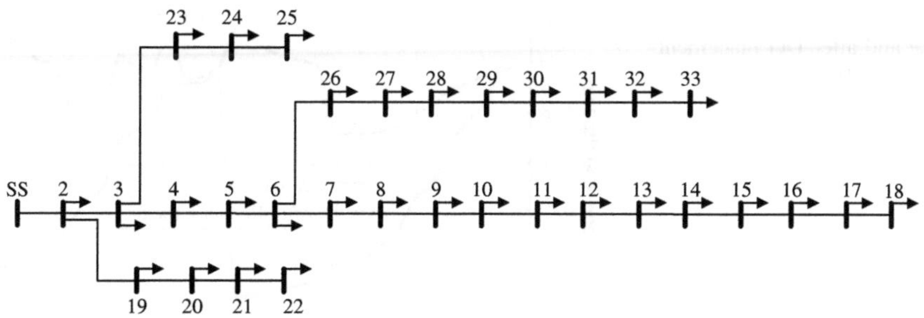

Fig. 3 IEEE 33 node test distribution system

5.1.1 Case 1

The procedure explained in the Sect. 4.1 is used and the system is as shown in Fig. 3. Initially, tie-switch placement is carried out at terminal nodes taking geographical constraints. The tie-switch locations selected are 18–22 and 25–33 and they will be in open position before reconfiguration. The DGs are present at the nodes 18 and 33. The DG sizing problem is done using *fmincon* solver using objective and constraints as given in Eqs. (9)–(13). The optimal real and reactive DG size obtained is given in Table 3.

The real and reactive power losses and real and reactive power generation fed from substation are given in Table 4. It can be observed that with reconfiguration algorithm considering both DG and tie-switches reduced the losses by 78 % and reliability also improved. The real and reactive power generation fed from substation after network reconfiguration is reduced by 59.4 %. Proper DG siting and sizing can have the benefit of voltage regulation which can seen from Fig. 4. The voltage profiles of the distribution network with proper DG siting and sizing considerably improved than in the base case.

The network reconfiguration with DG improved reliability with switch configuration 17–18 and 24–25 which is given in Table 5. The system down time is reduced to 9.2209 h/year from 15.39 h/year with the proposed method.

Table 3 Optimal size of DG for 33 node system

	Bus 18	Bus 33
P_{DG} (kW)	647	1143
Q_{DG} (kVar)	400	708

Table 4 Real and reactive power losses for 33 node system

	P_{loss} (kW)	Q_{loss} (kVAR)	P_{sub} (kW)	Q_{sub} (kVAR)
Base case	211	143	3925	2449
With DG	50	41	1974	1234
After reconfiguration	45	39	1591	639

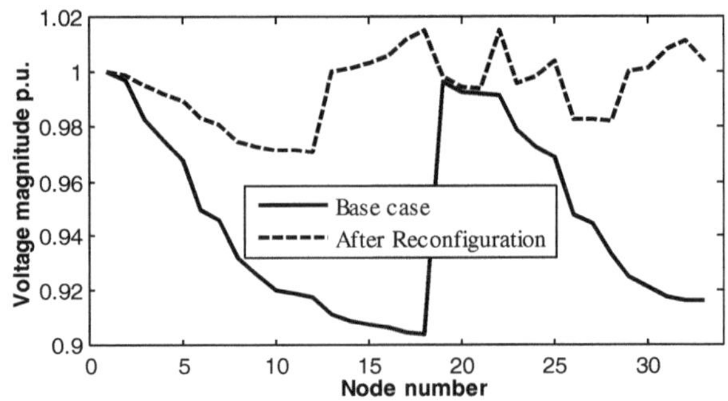

Fig. 4 Voltage profile for base and after DG placement for 33 node system

Table 5 Reliability output for 33 node distribution system

	Base case	After reconfiguration
Open switches	18–22, 33–25	17–18, 24–25
Unreliability (%)	0.18	7.1015e–002
Down time (h/year)	15.3961	9.2209

5.1.2 Case 2

The considered DG variables are location and size and fixed variables are type and number. The DG number is fixed to four and type is fixed to synchronous type. The load is taken as varying about 20 % increase and decrease with respect to normal load. The base case reliability indices SAIFI and SAIDI are 1.135 per customer and 14.618 h/customer, respectively. The objectives utilized in the formulation of optimal DG sizing and siting are the power loss reduction and reliability improvement. The problem formulation is solved using *fmincon* Matlab optimization tool which is given in Eqs. (14)–(26) and the Table 6 gives the results of the best DG locations and their optimal sizing for 35 % penetration level. It can be seen that, to minimize the loss, the optimal places of DGs are the low voltage buses whereas for improving only reliability, end feeder buses are the candidate buses for DGs placements.

For meeting all the objectives, the low voltage end feeder buses are the candidate buses for DGs placement. This shows the validity of the approach. It can be seen that these buses are the best choice for DGs placement to improve the voltage profile and reliability of the system. Considering all the objectives together, these buses or nearer buses are selected for the optimal DG locations. It should be noted that the sizes of these DGs are different for different cases. The voltage profiles with and without DGs corresponding to all the three cases along with base case are shown in Fig. 5. In the first two cases, only one objective at a time is emphasized,

Table 6 Optimal solution for the DG placement and size corresponding to 35 % penetration level and varying load (33-node system)

Penetration level = 35 %	Base case (without DG)	Case-1 (reliability improvement)	Case-2 (power loss reduction)	Case-3 (all objectives)
Optimal DG sizing (MW)	–	0.476, 0.010, 0.047, 0.767	0.378, 0.148, 0.378, 0.396	0.299, 0.170, 0.099, 0.731
Optimal DGs placement	–	18, 22, 25, 33	14, 16, 30, 32	13, 16, 18, 31
Power loss (kW)	211	59	51	52
SAIFI (interruptions/customer yr)	1.1351	0.0045	0.075953	0.0843
SAIDI (h/customer year)	14.618	0.108	0.510687	0.5580
Min voltage	0.898	0.9696	0.9663	0.9702

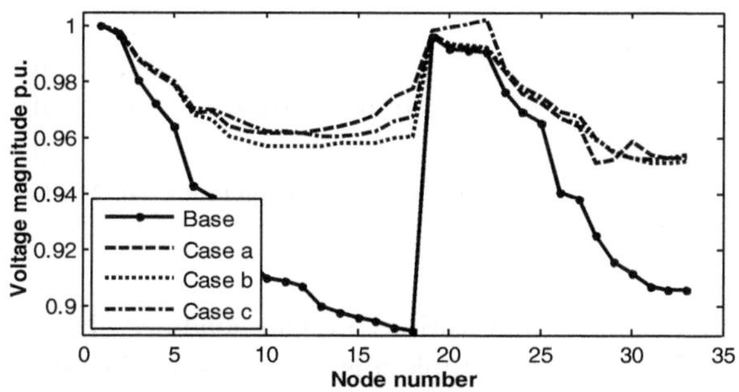

Fig. 5 Voltage profiles of 33 node system for base case and with DG placement for all three cases

which makes optimization tool to give more importance to that objective. In the third case, all the objectives are considered with weights assigned to them and so the result obtained is the contribution of all.

5.1.3 Case 3

Initially, tie-switches are placed in the system between buses 18–22 and 25–33, named as s1 and s2, respectively. There will be four possible binary tie-switch combinations and for each combination reconfiguration problem is carried out and the corresponding optimal switch combination and their respective objective values are given in Table 7.

It is clear that configuration-2 (C2) turn out to be the best configuration to operate the system without DG integration. DG placement is carried out for all the four configurations for various DG penetration levels, i.e., 20, 30, and 40 % of total

Table 7 Switch configuration after Reconfiguration for 33 node system

Configuration		Open switches	P_L (kW)	Q_{as} (%)
C1	Base case	18–22, 25–33	211	0.18
C2	s1, s2 close	7–8, 28–29	**83**	**0.15**
C3	s1 open and s2 close	25–33, 28–29	115	0.17
C4	s1 close and s2 open	7–8, 18–22	123	0.16

P_L Real power loss in the system in kW
Q_{as} Average system unavailability

Table 8 PVI for various configurations and DG penetration levels for 33 node system

C	DG penetration level								
	20 %			30 %			40 %		
	PVI (NL)	PVI (LI)	PVI (LD)	PVI (NL)	PVI (LI)	PVI (LD)	PVI (NL)	PVI (LI)	PVI (LD)
1	5.10	7.86	1.70	0.49	3.83	0.70	1.79	2.15	0.32
2	**0.30**	**0.46**	**0.09**	**0.07**	**0.22**	**0.01**	**0.06**	**0.10**	**0.01**
3	0.99	1.38	0.30	0.09	0.80	0.02	0.18	0.29	0.02
4	1.25	1.83	0.53	0.29	0.89	0.09	0.50	0.58	0.09

PVI Performance voltage index, *C* Configuration, *NL* Normal loading, *LI* Load increase, *LD* Load decrease

system load and various system loadings. For all possible DG combinations, optimal DG size problem is solved using Eqs. (9)–(13), power loss from Eq. (8), and reliability from Eqs. (1)–(5) by optimization tool. The DG placement results i.e., unreliability and power loss, for various configurations and penetration levels are now given to the AHP tool for obtaining the performance sensitivity from which the overall DG location can be obtained. The goal is to obtain the best DG locations and the objectives are loss minimization and reliability improvement and the alternatives are buses. The power loss and unavailability information at each bus for individual DG placement of each configuration are given as local priorities and the performance sensitivity (PS) from Eq. (27) is obtained from AHP tool. The above step is carried out for various DG penetration levels and various system loadings in order to obtain best DG location for any configuration are given in Table 8. The best configuration after DG placement is found using the existing performance voltage index (PVI) given in Eq. (28).

$$\mathrm{PVI} = \sum_{i=1}^{N} \frac{V^{\lim} - V_i}{V^{\max} - V^{\min}} \tag{28}$$

The resulting indices for various penetration levels and various configurations corresponding for different system loading conditions are given in Table 9. For low value of PVI, the best result is the obtained. From Tables 7 and 8, it is clear that configuration-2 (C2) is the optimal configuration to operate the system for the both with and without DG placement.

5.2 69-Bus Distribution System

The total real and reactive powers on the system are 3808.9 kW and 2691.4 kVar, respectively. In base case condition without DG, the total real power loss in the system is 247.5 kW. The maximum and the minimum bus voltage magnitudes, at base case condition, are 1.0 and 0.951 p.u., respectively, where the substation voltage (bus 1) is assumed to be 1.0 p.u.

5.2.1 Case 1

The tie-switch placement is carried out at end nodes. In this test system shown in Fig. 6 [31], the end nodes are 27, 35, 46, 50, 52, 65, 67, and 69. The tie-switches are placed at 27–35, 50–52, 46–67, and 65–69 and these will be in open position. The DG locations considered are 35, 50, 65, and 67. The DG sizing has been carried out using *fmincon* solver with loss minimization as objective and the constraints are meeting load, power balance equations, and voltage limits as in Eqs. (9)–(13). Table 10 gives the real and reactive optimal DG size for this test system.

The real and reactive power losses and real and reactive power generation fed from substation for the base case, with DG and after network reconfiguration with both DG and tie-switches are given in Table 11. The system losses reduce by 85 % with only DG and further reduce to 85 % with network reconfiguration. The voltage profiles under base case and after reconfiguration shown in Fig. 7 and the reliability output after network reconfiguration is given in Table 12 and the system down time reduced to 19.18 h/year from 22.27 h/year.

5.2.2 Case 2

For all predefined DGs configurations, the optimal size is found and the calculated objective values are stored for the calculation of the membership functions that are given in (14)–(26). The base case reliability indices SAIFI and SAIDI are 1.6369 per customer and 17.5294 h/customer, respectively. The fuzzy logic-based

Table 9 Overall best DG location by AHP for various DG penetration levels and their performance sensitivities by AHP for 33 node system

	Pl	DG locations for various configurations				OL	PS obtained from AHP
		C1	C2	C3	C4		
Normal loading	20 %	18	8	13	14	18	0.183
		22	22	15	22	14	0.136
		25	25	18	29	29	0.183
		33	33	33	31	33	0.128
	30 %	9	7	7	22	8	0.2
		18	22	15	25	15	0.15
		25	26	24	30	33	0.15
		33	33	33	33	31	0.145
	40 %	18	6	14	5	13	0.208
		25	8	7	25	18	0.167
		29	18	18	29	29	0.165
		33	33	33	31	14	0.125
20 % load increase	20 %	13	7	7	22	7	0.167
		16	17	19	25	17	0.167
		30	22	17	30	30	0.138
		33	23	33	32	33	0.138
	30 %	11	6	6	13	6	0.176
		14	22	13	25	25	0.059
		25	24	25	30	30	0.118
		30	33	33	33	33	0.176
	40 %	8	7	11	15	11	0.118
		18	22	24	22	22	0.235
		25	24	26	24	24	0.059
		30	33	33	33	33	0.059
20 % load decrease	20 %	13	20	12	22	16	0.107
		16	22	15	26	22	0.107
		30	27	18	31	18	0.143
		33	33	33	33	33	0.107
	30 %	13	7	8	14	8	0.152
		16	18	15	22	18	0.121
		18	22	18	30	22	0.152
		32	33	33	32	33	0.121
	40 %	10	6	8	18	8	0.115
		14	8	10	25	14	0.115
		30	22	16	25	25	0.154
		33	33	33	31	33	0.115

C1–C4 Configurations 1–4, *Pl* DG penetration level
OL Overall best DG location, *PS* Performance sensitivity

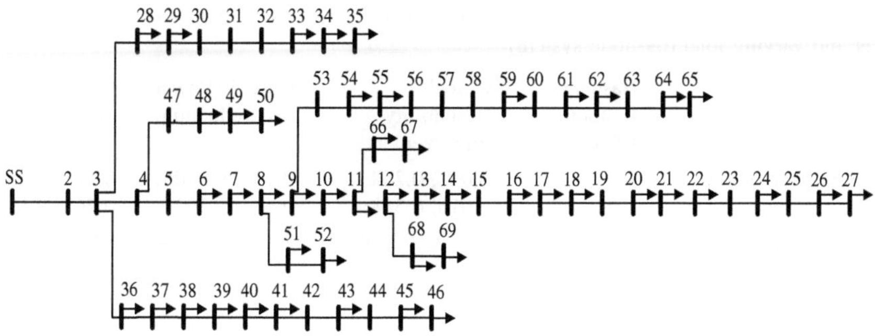

Fig. 6 IEEE 69 node test distribution system

Table 10 Optimal size of DG for 69 node system

	Bus 35	Bus 50	Bus 65	Bus 67
P_{DG} (kW)	297	556	1317	888
Q_{DG} (kVar)	210	394	933	629

Table 11 Real and reactive power losses for 69 node system

	P_{loss} (kW)	Q_{loss} (kVar)	P_{sub} (kW)	Q_{sub} (kVar)
Base case	256	116	4057	2810
With DG	49	23	793	550
After reconfiguration	36	17	293	198

Fig. 7 Voltage profile for base and after DG placement for 69 node system

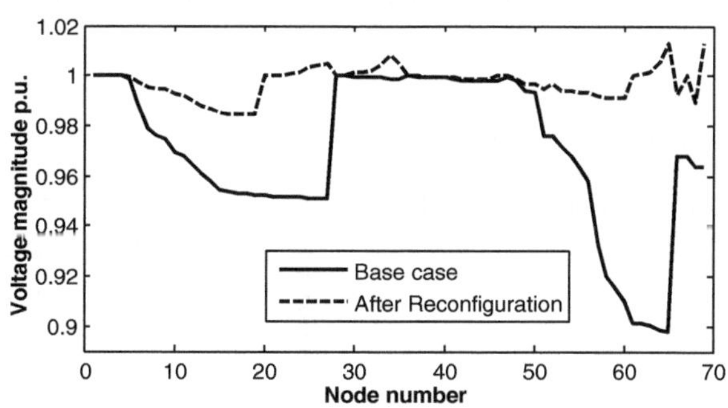

Table 12 Reliability output for optimal network configuration for 69 node distribution system

	Base case	After reconfiguration
Open switches	27–35, 50–52, 46–67, 65–67	19–20, 51–52, 45–46, 60–61
Unreliability (%)	0.25	0.22
Down time (h/year)	22.2704	19.1836

Table 13 Optimal solution for the DG placement and size corresponding to 35 % penetration level and varying load (69-node system)

Penetration level = 35 %	Base case (without DG)	Case-1 (Reliability improvement)	Case-2 (Power loss reduction)	Case-3 (All objectives)
Optimal DG sizing (MW)	–	0.037, 0.264, 0.010, 0.588	0.123, 0.0706, 0.644, 0.302	0.121, 0.080, 0.768, 0.181
Optimal DGs placement	–	18, 24, 52, 65	24, 27, 61, 64	24, 27, 61, 65
Power loss (kW)	256	76	58	59
SAIFI (interruptions/customer yr)	1.636	0.232	0.265	0.261
SAIDI (h/customer yr)	17.529	1.50	1.517	1.561
Min voltage	0.897	0.959	0.9563	0.969

min-max problem is used for finding the best DG placement. Table 13 shows the results of the best DGs location and their optimal sizing for 35 % penetration levels. In this system, there are eight terminating nodes and low voltages at buses 27 and 65. It can be seen that these buses are the best choice for DGs placement to improve the voltage profile and reliability of the system.

Considering all the objectives together, these buses or nearer buses are selected for the optimal DG locations. It should be noted that the sizes of these DGs are different for different objectives. The voltage profiles with and without DGs corresponding to all the cases are shown in Fig. 8 for the test system. It can be noted that the DGs sizes are different for each case for both test systems. In the case a and case b, only one objective at a time is emphasized, which makes optimization tool to give more importance to that objective. In the case c, all the objectives are

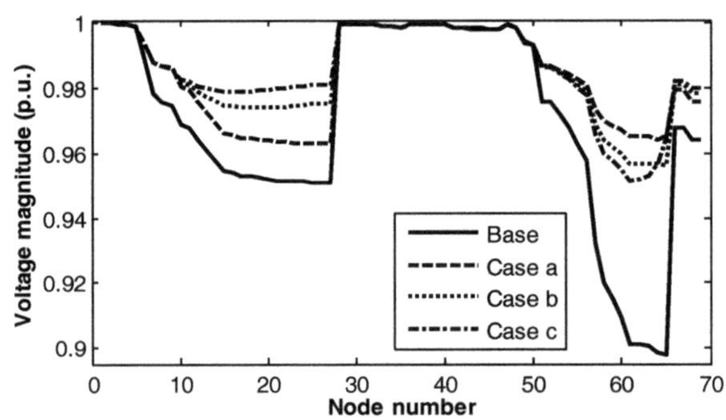

Fig. 8 Voltage profiles of 69 node system for base case and with DG placement for all three cases

considered with weights assigned to them and so the result obtained is the contribution of all.

5.2.3 Case 3

Initially, tie-switches are placed in the system between 27–35, 50–52, 46–67, and 65–69 and named s1, s2, s3, and s4, respectively. There will be sixteen possible tie-switch binary (open/close) combinations. The optimal switch configuration considering power loss reduction and loss minimization as the objective is obtained by the reconfiguration algorithm for each tie-switch set using the algorithm given in proposed methodology and the obtained optimal configurations are given in Table 14. Now, DG placement is carried out for all the configurations. The combinations of DG are generated and the DG number is limited to four. For each DG combination, optimal DG size from Eqs. (9)–(13), power loss from Eq. (8), and system reliability from Eqs. (1)–(5) are found using the search algorithm as shown in Fig. 4. This is repeated for all configurations and for various DG penetration levels, i.e., 20, 30 and 40 % of total system load. The above step is repeated for different loading conditions, i.e., 20 % load increase and 20 % load decrease. The overall best DG location for any configuration for various penetration levels and the detailed PS given in Eq. (27) is used for the overall best locations is obtained from the AHP tool are given in Table 15. The down time in h/year, for configurations

Table 14 Switch configuration after reconfiguration for 69 node system

Configuration		Open switches	PL	Qas
C1	Base case	27–35, 50–52, 45–67, 65–69	211	0.0025
C2	s1, s2, s3, s4 close	15–16, 41–42, 60–61, 48–49	**42**	**0.0015**
C3	s2, s3, s4 close	49–49, 11–66, 60–61	115	0.0017
C4	s1, s2, s4 close	15–16, 48–49, 60–61	123	0.0016
C5	s1, s3, s4 close	15–16, 41–42, 60–61	103	0.0018
C6	s1, s2, s3 close	11–66, 48–49, 15–16	220	0.0019
C7	s4 close	60–61	50	0.0023
C8	s3 close	11–66	253	0.0025
C9	s3, s4 close	11–66, 60–61	59	0.0022
C10	s2 close	48–49	253	0.0025
C11	s1, s3 close	15–16, 11–66	115	0.0017
C12	s2, s3 close	48–49, 11–66	250	0.0025
C13	s1 close	15–16	225	0.0020
C14	s1, s4 close	15–16, 60–61	65	0.0017
C15	s2, s4 close	15–16, 11–66	115	0.0017
C16	s1, s2 close	15–16, 48–49	223	0.0021

P_L Real power loss in the system in kW
Q_{as} Average system unavailability

Table 15 Overall best DG location by AHP from various configurations and DG penetration levels for 69 node system

	Configurations	DG penetration level		
		20 %	30 %	40 %
Normal loading	C1	27, 59, 62, 65	25, 63, 65, 69	27, 57, 61, 65
	C2	9, 13, 27, 69	8, 21, 46, 69	9, 11, 50, 69
	C3	11, 15, 27, 69	9, 18, 67, 69	11, 17, 50, 69
	C4	9, 11, 27, 69	9, 22, 35, 69	8, 20, 50, 69
	C5	7, 11, 50, 69	9, 49, 67, 69	7, 11, 50, 69
	C6	15, 50, 61, 65	17, 27, 63, 65	50, 55, 61, 65
	C7	11, 23, 50, 69	11, 21, 50, 69	10, 17, 27, 69
	C8	43, 63, 65, 67	27, 50, 62, 64	14, 27, 61, 65
	C9	12, 18, 46, 69	16, 50, 54, 69	9, 19, 50, 69
	C10	52, 61, 63, 67	59, 61, 65, 69	17, 61, 65, 67
	C11	61, 63, 65, 67	35, 58, 63, 65	16, 35, 61, 65
	C12	50, 58, 61, 65	27, 60, 63, 65	50, 52, 61, 63
	C13	61, 63, 65, 69	61, 63, 65, 69	52, 61, 64, 69
	C14	11, 50, 59, 69	25, 50, 66, 69	11, 50, 59, 69
	C15	11, 18, 27, 69	9, 17, 27, 69	17, 51, 67, 69
	C16	52, 61, 63, 65	35, 59, 61, 65	52, 61, 63, 65
	Overall best DG location	11, 65, 18, 69	18, 21, 69, 65	9, 61, 27, 69
	PS	0.146, 0.098, 0.098, 0.073	0.098, 0.098, 0.073, 0.073	0.138, 0.098, 0.069, 0.069
20 % load increase	C1	52, 55, 62, 65	17, 27, 61, 65	24, 52, 61, 65
	C2	11, 35, 55, 69	8, 24, 67, 69	8, 23, 50, 69
	C3	15, 27, 58, 69	11, 19, 52, 69	9, 19, 50, 69
	C4	25, 59, 67, 69	21, 55, 67, 69	17, 51, 67, 69
	C5	11, 27, 58, 69	10, 50, 58, 69	9, 19, 50, 69
	C6	52, 57, 62, 65	10, 61, 65, 67	27, 56, 61, 65
	C7	18, 20, 67, 65	22, 54, 67, 69	17, 49, 67, 69
	C8	60, 63, 65, 67	7, 27, 62, 65	19, 27, 61, 65
	C9	18, 27, 59, 69	11, 22, 50, 69	10, 21, 50, 69
	C10	27, 42, 61, 65	25, 61, 65, 67	13, 27, 61, 65
	C11	31, 46, 61, 65	8, 62, 65, 67	35, 45, 61, 65
	C12	27, 61, 63, 65	23, 35, 62, 65	23, 27, 61, 65
	C13	39, 52, 61, 65	52, 58, 61, 65	5, 61, 65, 67
	C14	27, 58, 66, 69	10, 50, 58, 69	24, 50, 55, 69
	C15	21, 27, 58, 69	18, 56, 67, 69	9, 22, 50, 69
	C16	18, 52, 62, 65	13, 50, 62, 65	61, 63, 67, 69
	Overall best DG location	27, 52, 65, 69	27, 22, 67, 69	17, 61, 65, 69
	PS	0.032, 0.012, 0.013, 0.019	0.031, 0.025, 0.019, 0.025	0.024, 0.012, 0.024, 0.018

(continued)

Table 15 (continued)

	Configurations	DG penetration level		
		20 %	30 %	40 %
20 % load decrease	C1	27, 59, 61, 65	59, 61, 65, 67	21, 61, 65, 69
	C2	10, 50, 64, 69	9, 50, 64, 67	4, 11, 50, 69
	C3	11, 16, 50, 69	18, 50, 54, 69	4, 23, 52, 69
	C4	11, 16, 50, 69	10, 50, 61, 69	10, 16, 50, 69
	C5	10, 12, 50, 69	11, 50, 52, 64	11, 29, 50, 65
	C6	27, 60, 62, 65	50, 53, 61, 65	52, 59, 61, 69
	C7	12, 18, 50, 69	7, 17, 50, 69	11, 20, 50, 65
	C8	50, 57, 63, 65	23, 52, 61, 65	27, 52, 62, 64
	C9	9, 15, 50, 69	8, 20, 50, 69	7, 16, 50, 69
	C10	46, 61, 63, 65	27, 41, 61, 65	29, 61, 65, 69
	C11	52, 59, 61, 65	61, 63, 65, 69	50, 52, 61, 63
	C12	17, 61, 65, 67	61, 63, 65, 67	57, 61, 65, 69
	C13	20, 35, 63, 65	50, 61, 63, 65	60, 62, 65, 69
	C14	11, 50, 58, 69	12, 49, 65, 67	4, 7, 50, 69
	C15	11, 17, 46, 65	3, 7, 65, 67	20, 28, 52, 69
	C16	24, 27, 62, 65	61, 63, 65, 67	61, 63, 65, 69
	Overall best DG location	27, 57, 18, 65	18, 50, 65, 67	52, 50, 65, 69
	PS	0.036, 0.045, 0.030, 0.030	0.022, 0.014, 0.036, 0.022	0.014, 0.016, 0.029, 0.022

C1–C4 Configurations 1–4, *Pl* DG penetration level
OL Overall best DG location, *PS* Performance sensitivity

1–16 and for the overall best DG placement are 15.62, 12.454, 15.836, 12.440, 11.79, 14.38, 15.26, 15.27, 14.90, 15.54, 12.37, 15.18, 12.73, 13.42, 16.19, 14.74, h/year, respectively, calculated for any DG penetration level. The performance voltage index for all the configurations is given in Table 16. The best configuration to operate the system, with and without DG integration is found from the Tables 14 and 16 and is came out to be configuration-2 (C2). This is also carried out for various system loadings, i.e., 20 % load increase and 20 % load decrease. In the same way, the best DG location for any configuration corresponding to various DG penetration levels is also found.

Table 16 PVI for various configurations and DG penetration levels for 69 node system

C	DG penetration level								
	20 %			30 %			40 %		
	PVI (NL)	PVI (LI)	PVI (LD)	PVI (NL)	PVI (LI)	PVI (LD)	PVI (NL)	PVI (LI)	PVI (LD)
1	0.37	7.71	2.53	0.26	4.59	1.41	0.26	1.77	0.17
2	**0.10**	**0.02**	**0.01**	**0.08**	**0.01**	**0.01**	**0.11**	**0.01**	**0.01**
3	0.11	0.16	0.02	0.09	0.03	0.01	0.11	0.06	0.02
4	0.09	0.02	0.05	0.10	0.01	0.02	0.09	0.02	0.02
5	0.10	0.02	0.09	0.09	0.03	0.07	0.11	0.03	0.04
6	0.29	2.92	0.15	0.25	1.25	0.32	0.21	0.30	0.04
7	0.09	0.16	0.02	0.08	0.02	0.01	0.12	0.06	0.01
8	0.37	7.29	2.48	0.32	3.74	0.98	0.27	1.84	0.19
9	0.09	0.15	0.02	0.10	0.10	0.01	0.10	0.12	0.01
10	0.38	7.56	2.72	0.33	4.61	0.96	0.28	2.10	0.31
11	0.29	3.01	1.52	0.25	1.14	0.30	0.22	0.30	0.09
12	0.37	7.02	2.37	0.33	3.98	0.98	0.26	1.78	0.31
13	0.29	3.11	1.51	0.24	1.20	0.36	0.18	0.41	0.03
14	0.10	0.03	0.01	0.09	0.04	0.01	0.10	0.02	0.01
15	0.10	0.12	0.03	0.11	0.01	0.02	0.10	0.02	0.01
16	0.29	2.98	1.49	0.25	1.19	0.42	0.21	0.31	0.02

PVI Performance voltage index, *C* Configuration, *NL* Normal loading, *LI* Load increase, *LD* Load decrease

6 Conclusions

The reliability and power quality studies are carried out in three different ways. The renewable DG technologies including the synchronous machine-based biomass generator was considered in distribution system expansion planning.

- The network reconfiguration with DGs and tie-switch placement for the reliability improvement and loss minimization is proposed. Tie-switches are placed at terminal nodes with geographical constraints and their all possible binary combinations are considered. DGs sizing for the predefined DGs are calculated using integer programming for objectives reliability improvement and minimization of losses with penetration level taken for less than total load. Power loss is calculated using forward/backward sweep distribution load flow algorithm in all the three cases. A search-based reconfiguration algorithm has been formulated for finding the optimal switch configuration for the radial distribution system with DG. It can be observed from the results that the reductions in the real power loss from the base case are 78 and 85 % for the 33-bus and

69-bus distribution systems, respectively. The improvements in the system reliability, obtained using the proposed method, are 40 and 14 % for the 33-bus and 69-bus distribution systems, respectively.

- The optimal DG placement and DG sizing for combined objectives of reliability improvement and loss minimization, using fuzzy logic approach is proposed in two steps. Out of four DG variables, type, and number are taken as fixed variables and size and location are considered as variables in the formulation. In the first step, for all DGs combination, their optimal sizes are calculated using nonlinear optimization tool with the improved reliability and real power loss reduction objectives. SAIDI and SAIFI indices are used for the reliability evaluation in this case. In the formulation three indices, power loss, SAIDI and SAIFI are used with weighing factors and taken as combined objective. Three cases for different weights assigned to the objectives are solved for the DGs placement and sizing which is carried out for 35 % DG penetration level. Fuzzy min-max problem is applied to the multi-objectives for finding the optimal DG locations. It is observed from the results that the reduction in the real power loss from the base case is 75 and 77 % for the 33-bus and 69-bus distribution systems, respectively. The improvements in the reliability at each load point obtained, using the proposed method, are 96 and 91 % for the 33-bus and 69-bus distribution systems, respectively. The minimum voltage of 0.9702 p.u. is obtained which is much better than the base case which is 0.898 p.u.
- A combined network reconfiguration and optimal DG siting and sizing is carried out for the objectives real power loss reduction and reliability improvement, for various penetration levels and various loading conditions. It is found from the results that the reduction in the real power loss over the base case is 60 and 80 % for the 33-bus and 69-bus distribution systems, respectively. The improvement in the reliability obtained using the proposed method is 16 and 40 % for the 33-bus and 69-bus distribution systems, respectively. From studied results, it is also observed that reliability and loss of the network are mainly depends on power demand of the network, type and capacity of the distributed generations and their location in the network. The network condition has huge impact on DG allocation. The configuration-2 comes out to be the best configuration of the system to operate.

It is observed that optimal sitting and sizing of the DGs improve the system performance in terms of real power loss reduction and improvement in the system reliability. Different objectives give different optimal locations and sizes but based on the physical constraints, the final decisions are to be made. The present work may be a guideline to the planners and policy makers for planning the DGs in the distribution systems.

References

1. Gonen T (1986) Electric power distribution system. McGraw Hill, Inc.
2. Billinton R, Allan RN (1996) Reliability evaluation of power systems, 2nd edn. Plenum, New York
3. Borges CLT, Falcao DM (2006) Optimal distributed generation allocation for reliability, losses and voltage improvement. Int J Electr Power Energy Syst 28(6):413–420
4. Soroudi A, Afrasiab M (2012) Binary PSO-based dynamic multi-objective model for distributed generation planning under uncertainty. IET Renew Power Gener 6(2):67–78
5. Moradi MH, Abedini M (2012) A combination of genetic algorithm and particle swarm optimization for optimal DG location and sizing in distribution systems. Int J Electr Power Energy Syst 34(1):66–74
6. Ettehadi M, Ghasemi H, Vaez-Zadeh S (2013) Voltage stability-based DG placement in distribution networks. IEEE Trans Power Del 28(1):171–178
7. Amanulla B, Chakrabarti S, Singh SN (2012) Reconfiguration of power distribution systems considering reliability and power loss. IEEE Trans Power Del 27(2):918–926
8. Mao Y, Miu KN (2003) Switch placement to improve system reliability for radial distribution systems with distributed generation. IEEE Trans Power Syst 18(4):1346–1352
9. Han L, Zhou R, Deng X (2009) An analytical method for DG placements considering reliability improvements. In: IEEE PES GM, July 2009, pp 1–5
10. Rao RS, Narasimham SVL, Raju MR, Rao AR (2011) Optimal network reconfiguration of large-scale distribution system using harmony search algorithm. IEEE Trans Power Syst 26 (3):1080–1088
11. Rao RS, Ravindra K, Satish K, Narasimham SVL (2013) Power loss minimization in distribution system using network reconfiguration in the presence of distributed generation. IEEE Trans Power Syst 28(1):317–325
12. Georgilakis PS, Hatziargyriou ND (2013) Optimal distributed generation placement in power distribution networks: models, methods, and future research. IEEE Trans Power Syst 28 (3):3420–3428
13. IEEE Application Guide for IEEE Std 1547, IEEE Standard for Interconnecting Distributed Resources With Electric Power Systems 2008
14. Walling RA, Saint R, Dugan RC, Burke J, Kojovic LA (2008) Summary of distributed resources impact on power delivery systems. IEEE Trans Power Del 23(3):1636–1644
15. Wang C, Hehrir MH (2004) Analytical approaches for optimal placement of distributed generation sources in power systems. IEEE Trans Power Syst 19(4):2068–2076
16. Acharya N, Mahat P, Mithulananthan N (2006) An analytical approach for DG allocation in primary distribution network. Int J Electr Power Energy Syst 28(10):669–678
17. Ettehadi M, Ghasemi H, Vaez-Zadeh S (2013) Voltage stability-based DG placement in distribution networks. IEEE Trans Power Del 28(1):171–178
18. Al Abri RS, El-Saadany EF, Atwa YM (2013) Optimal placement and sizing method to improve the voltage stability margin in a distribution system using distributed generation. IEEE Trans Power Syst 28(1):326–334
19. Esmaili M (2013) Placement of minimum distribution generation units observing power losses and voltage stability with network constraints. IET Gen Transm Distrib 7(8):813–821
20. Lee SH, Park J-W (2003) Optimal placement and sizing of multiple DGs in a practical distribution system by considering power loss. IEEE Trans Ind Appl 49(5):2262–2270
21. Kroposki B, Sen PK, Malmedal K (2003) Optimum sizing and placement of distributed and renewable energy sources in electric power distribution systems. IEEE Trans Ind Appl 49 (6):2741–2752
22. Pandi VR, Zeineldin HH, Xiao W (2013) Determining optimal location and size of distributed generation resources considering harmonic and protection coordination limits. IEEE Trans Power Syst 28(2)1245–1254

23. Hien NC, Mithulananthan N, Bansal RC (2013) Location and sizing of distributed generation units for loadability enhancement in primary feeder. IEEE Syst J 7(4):797–806
24. Tan WS, Hassan MY, Rahman HA, Abdullah Md P, Hussin F (2013) Multi-distributed generation planning using hybrid particle swarm optimisation- gravitational search algorithm including voltage rise issue. IET Gen Transm Distrib 7(9):929–942
25. Soroudi A, Ehsan M, Caire R, Hadjsaid N (2011) Possibilistic evaluation of distributed generations impacts on distribution networks. IEEE Trans Power Syst 26(4):2293–2301
26. Das D (2006) A fuzzy multiobjective approach for network reconfiguration of distribution systems. IEEE Trans Power Del 21(1):202–209
27. Zimmermann HJ (1999) Fuzzy programming and linear programming with several objectives. Proc Inst Elect Eng Gen Transm Distrib 641–648
28. Pavani P, Singh SN (2013) Reconfiguration of radial distribution networks with distributed generation for reliability improvement and loss minimization. In: IEEE PES GM Conference, Vancouver, Canada, July 2013
29. Alessio I, Ashraf L (2009) Analytic hierarchy process and expert choice: benefits and limitations. OR Insight 22(4):201–220
30. Baran ME, Wu F (1989) Network reconfiguration in distribution system for loss reduction and load balancing. IEEE Trans Power Del 4(2):1401–1407
31. Savier JS, Das D (2007) Impact of network reconfiguration on loss allocation of radial distribution systems. IEEE Trans Power Del 2(4):2473–2480

Chapter 5
Tracing the Unreliability and Recognizing the Weak Parts of a Power System

Kaigui Xie

Abstract This chapter titled "Tracing the Unreliability and Recognizing the Weak Parts of a Power System" presents two unreliability tracing principles, i.e., failure component sharing principle (FCSP) and proportional sharing principle (PSP) to derive the UT model and algorithm for a power system. The proposed technique provides a new idea for evaluating the impact of components on the system unreliability and can be used to distribute the unreliability indices of a power system to the system components impartially and reasonably. The models and algorithms for tracing the unreliability are determined using the PSP. The UT indices for individual component are created to recognize the major unreliability contributions (MUC) of the power system. The method provides the system planner/operator with information on how much an individual component contributes to the overall system risk and the system MUC. The basic principles and methods proposed in this paper are general and therefore applicable to both HL-I and HL-II. The illustration application is at HL-I and HL-II, and the RBTS and IEEE-RTS were examined using the proposed method. The results show that the proposed approach has good practicability and has promising prospects in other engineering applications. Power system applications utilizing the proposed UT technique can be used to recognize weak system elements, improve the reliability, analyze reliability cost/benefits, and optimize scheduled maintenance. In addition, the proposed UT technique can be easily implemented on the existing reliability evaluation programs with a relatively little additional computation requirement, and conveniently expanded to AC/DC systems, distribution systems or interconnected systems.

K. Xie (✉)
School of Electrical Engineering, Chongqing University, Chongqing, China
e-mail: kaiguixie@vip.163.com

N.R. Karki et al. (eds.), *Sustainable Power Systems*, Reliable and Sustainable Electric Power and Energy Systems Management,
DOI 10.1007/978-981-10-2230-2_5

1 Introduction

The Roy Billinton Test System (RBTS) is a small reliability test system shown in Fig. 1 [1]. The reliability performance of individual components has different impacts on the system reliability and it is both interesting and important to consider. How much is the "contribution" of a selected component to the system unreliability and what are the major unreliability contributions (MUC) in the RBTS? This chapter introduces an unreliability tracing (UT) technique to answer these questions. The UT approach allows the unreliability performance of a power system expressed by basic reliability indices [2, 3], such as the loss of load probability (LOLP), loss of load frequency (LOLF) and expected energy not supplied (EENS), to be traced or allocated to every component in the power system.

A quantitative assessment of the "contribution" of a selected component to the system unreliability will be useful to identify the system weak parts (WP), and to compare those using quantitative indices. This information can provide the power system planners and operators with useful information in reliability assessment for power system planning and operation.

Considerable work has been done [1–10] in developing power system reliability models and algorithms, and reliability application and risk analysis, but there has been relatively little research on UT techniques. The relevant research is mainly focused on reliability analysis considering the class, order of contingencies, and the component failure causes.

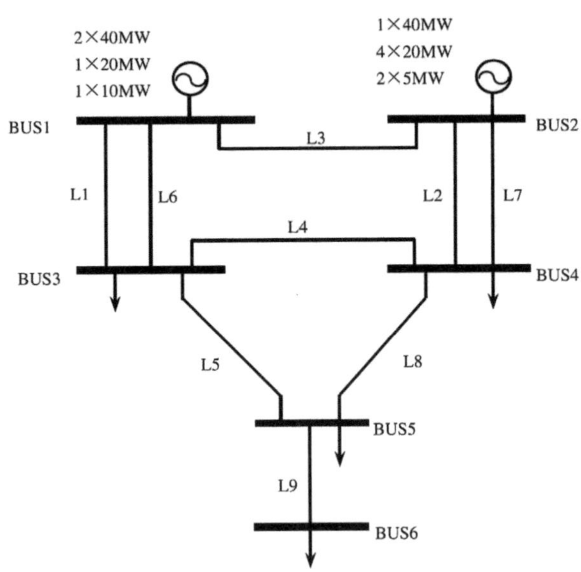

Fig. 1 Single line diagram of the RBTS

Relative contributions to the overall system indices from composite generation and transmission failures and distribution zone failures are illustrated in [4, 5]. References [6, 7] present the impact proportions of the generation, line and transformer classes on the system reliability. A number of important reliability influence factors, such as adverse weather, defective equipment, tree contact, loss of supply, scheduled outage, foreign interference, lighting, etc., are illustrated and analyzed using historical reliability performance data from a number of Canadian utilities [8]. These papers give the unreliability contribution of the different classes or contingency levels, but do not give the unreliability contribution from each component.

References [9, 10] propose a method for allocating the transmission reliability cost to generators and the transmission lines based on their reliability contributions. These two papers give the cost contribution of different components, but do not give the reliability contribution of each component.

References [11, 12] present methods to calculate the sensitivity of reliability indices with respect to component reliability parameters. References [13, 14] study the impact of component reliability parameters on the system reliability, and the changes in power system reliability indices with different component reliability parameters. It is possible to recognize the WP of a power system using the sensitivity studies. The above works, however, do not provide the unreliability contributions of individual system components.

The UT technique has been applied to the reliability study of generation systems and bulk power system (BPS) [15–17]. The failure component sharing principle (FCSP) and proportional sharing principle (PSP) are proposed to derive the UT model and the algorithm for power systems. The applications of the proposed UT technique to analyze the unreliability contribution of power system components and to recognize the WPs are illustrated in this chapter.

2 Unreliability Tracing Principles

There are many system states analyzed in a power system reliability evaluation process. The contributions to reliability indices, such as the LOLP, LOLF and EENS, etc., can be evaluated using the relevant reliability assessment technique in each sampled or enumerated state [2–4]. In every state, other than the base case, there are one or more component failures.

For example, assume that in a BPS, a state occurs with an LOLP = 0.00234, in which generating unit 2, transformer 3, and line 11 are failed and the remaining components are healthy. In this state, the LOLP results from the three failure components. The question remains, how is the LOLP attributed to each component?

The purpose of using the UT technique is to determine the allocation proportion of these failed components to the unreliability indices associated with a given state, and to evaluate how much an individual component contributes to the overall system unreliability indices. This chapter presents the following two UT principles, the FCSP and the PSP:

(1) Failure component sharing principle (FCSP): Only the failed components take on the responsibility for the system unreliability. In other words, the healthy components are not considered in the unreliability tracing.
(2) Proportional sharing principle (PSP): The reliability indices are proportionally distributed among all the system components according to the PSP described in [15, 16].

Assume a system consisting of N components C_i ($i = 1, 2, \ldots, N$). Let P_i ($i = 1, 2, \ldots, N$) represent the performance parameters, which are related only to the component i. Let $f(P_1, P_2, \ldots, P_N)$ represent a reliability index of a state, such as LOLP, LOLF and EENS, which is the function of P_1, P_2, …, and P_N.

For simplicity, supposing a failure state S results from the outage of C_1 and C_2. It is assumed that $f(P_1, P_2, \ldots, P_N)$ can be divided into three unconnected terms where the first term $f_1(P_1)$ is only related to C_1, the second term $f_2(P_2)$ is only related to C_2, and the third term $g(P_3, P_4, \ldots, P_N)$ is related to the other components. So

$$f(P_1,P_2,\ldots,P_N) = f_1(P_1)f_2(P_2)g(P_3,P_4,\ldots,P_N) \tag{1}$$

The obligation for a failure system state can be assigned to the failed components, i.e., C_1 and C_2. From (1), the terms $f_{(S\to 1)}$ and $f_{(S\to 2)}$ can be used to demonstrate the PSP and describe the unreliability obligation of C_1 and C_2, respectively:

$$f_{(S\to 1)} = \frac{f_1(P_1)}{f_1(P_1)+f_2(P_2)} f(P_1,P_2,\ldots,P_N) \tag{2}$$

$$f_{(S\to 2)} = \frac{f_2(P_2)}{f_1(P_1)+f_2(P_2)} f(P_1,P_2,\ldots,P_N), \tag{3}$$

where the subscripts 1 and 2 are the component number.

It can be seen from (2) and (3) that this principle has two features:

1. The PSP has symmetry in which the P_1 (P_2) is only a symbol and it can be used to represent any component. If the component calculating positions are exchanged, the tracing results will not change.
2. The PSP has an identity. This means that

$$f_{(S\to 1)} + f_{(S\to 2)} \equiv f(P_1,P_2,\ldots,P_N) \tag{4}$$

This principle basically assumes that the system is a perfect 'mixer' of every term so that it is impossible to tell which particular term will go into some particular term. As it is impossible to 'dye' the reliability index and check the color of the out terms, the PSP can be neither proved nor disproved, which is the same as the PSP used to trace the power flow [18]. In this respect, the principle is fair as it treats all terms in the same way. In other words, no particular term is distinguished in any way.

3 Tracing the Probability and Frequency of a System Failure State

3.1 Tracing the Probability of a System Failure State

In order to simplify the derivation process, it is assumed that the component failures are exponentially distributed and each component resides in two states, i.e., the up state and the down state. Using the FCSP and PSP and Eqs. (2) and (3), the proposed UT technique can be extended to trace the unreliability contributions of a power system considering system components that can reside in more than two states.

The probability of a system failure state S can be given by [2, 3, 19]

$$P_{(S)} = \prod_{i \in S_D} U_i \times \prod_{j \in S_U} (1 - U_j) \tag{5}$$

where U_i and U_j are the unavailability of the ith and jth component, respectively; S_D and S_U are the component set of down states and up states, respectively.

According to Eqs. (2), (3) and the UT principles, the probability tracing results of the system failure state S is given by

$$\begin{cases} P_{(S \to i)} = P_{(S)} \times \frac{U_i}{\sum\limits_{j \in S_D} U_j} & i \in S_D \\ P_{(S \to i)} = 0 & i \in S_U \end{cases}, \tag{6}$$

where $P_{(S \to i)}$ is the probability distribution results of component i.

Equation (7) indicates the identity feature of the PSP in Eq. (4).

$$P_{(S)} \equiv \sum_{i \in S_D} P_{(S \to i)} + \sum_{i \in S_U} P_{(S \to i)}. \tag{7}$$

Let a system failure state S with two failed components, i.e. C_1 and C_2, as an example. Based on the reliability evaluation principle, the LOLP of state S with C_1 and C_2 failed is:

$$\begin{aligned} \mathrm{LOLP}_S &= f(P_1, P_2, \ldots, P_N) \\ &= \frac{\lambda_1}{\lambda_1+\mu_1} \times \frac{\lambda_2}{\lambda_2+\mu_2} \times \frac{\mu_3}{\lambda_3+\mu_3} \cdots \frac{\mu_N}{\lambda_N+\mu_N} \\ &\triangleq U_1 \times U_2 \times (1-U_3)\ldots(1-U_N) \end{aligned} \tag{8}$$

In (8), P_1, P_2 are only related to C_1 and C_2, respectively. Based on the PSP, i.e. (2) and (3), the tracing results of the reliability index LOLP are given by:

$$\mathrm{LOLP}_{(S\to 1)} = \frac{U_1}{U_1+U_2} \times \mathrm{LOLP}_S \tag{9}$$

$$\mathrm{LOLP}_{(S\to 2)} = \frac{U_2}{U_1+U_2} \times \mathrm{LOLP}_S. \tag{10}$$

3.2 *Tracing the Frequency of a System Failure State*

The frequency of a system failure state S can be calculated by [2, 3, 19]

$$F_{(S)} = P_{(S)}\left(\sum_{i\in S_D} \mu_i + \sum_{j\in S_U} \lambda_j\right), \tag{11}$$

where μ_i and λ_j are the repair rate and failure rate of the ith and jth component, respectively.

Assume that the sequence number of the components in set S_D is 1, 2, …, n. Equation (11) can be rewritten as

$$\begin{aligned} F_{(S)} = &\prod_{i\in S_D} U_i \times \prod_{j\in S_U} (1-U_j) \times \mu_1 + \prod_{i\in S_D} U_i \times \prod_{j\in S_U} (1-U_j) \times \mu_2 + \cdots \\ &+ \prod_{i\in S_D} U_i \times \prod_{j\in S_U} (1-U_j) \times \mu_i + \prod_{i\in S_D} U_i \times \prod_{j\in S_U} (1-U_j) \times \sum_{j\in S_U} \lambda_j \end{aligned} \tag{12}$$

A calculation process similar to that in Eq. (6) can be used to trace the failure frequency of the failure state S

$$\begin{cases} F_{(S\to i)} = & P_{(S)} \times \mu_1 \times \frac{U_i}{U_1\mu_1 + \sum\limits_{j\in S_D \text{ and } j\neq 1} U_j} \\ & + P_{(S)} \times \mu_2 \times \frac{U_i}{U_2\mu_2 + \sum\limits_{j\in S_D \text{ and } j\neq 2} U_j} + \cdots \\ & + P_{(S)} \times \mu_i \times \frac{U_i}{U_i\mu_i + \sum\limits_{j\in S_D \text{ and } j\neq i} U_j} & i \in S_D \\ & + \ldots + P_{(S)} \times \mu_n \times \frac{U_i}{U_n\mu_n + \sum\limits_{j\in S_D \text{ and } j\neq n} U_j} \\ & + P_{(S)} \times \sum\limits_{j\in S_D} \lambda_j \times \frac{U_i}{\sum\limits_{j\in S_D} U_j} \\ F_{(S\to i)} = 0 & & i \in S_U \end{cases}, \tag{13}$$

where $F_{(S\to i)}$ is the frequency distribution results of component i.

Similarly, Eq. (14) also proves the identity feature of the PSP.

$$F_{(S)} \equiv \sum_{i\in S_D} F_{(S\to i)} + \sum_{i\in S_U} F_{(S\to i)}. \tag{14}$$

Equations (7) and (14) indicate that the sum of the probability and frequency distribution results is equal to the original value. The probability distribution results of a system failure state are only related to the unavailability of individual components, whereas the frequency distribution results are also related to the component transition rates.

Let a system failure state with two failed components, i.e. C_1 and C_2, as an example. Based on the PSP, the tracing results of the reliability index LOLF are:

$$\text{LOLF}_{(S\to 1)} = \frac{U_1\mu_1}{U_1\mu_1 + U_2} U_1\mu_1 U_2 X_1 + \frac{U_1}{U_1 + U_2\mu_2} U_1 U_2 \mu_2 X_1 + \frac{U_1}{U_1 + U_2} U_1 U_2 X_2 \tag{15}$$

$$\text{LOLF}_{(S\to 2)} = \frac{U_2}{U_1\mu_1 + U_2} U_1\mu_1 U_2 X_1 + \frac{U_1\mu_2}{U_1 + U_2\mu_2} U_1 U_2 \mu_2 X_1 + \frac{U_2}{U_1 + U_2} U_1 U_2 X_2. \tag{16}$$

where $X_1 = A_3A_4A_N; X_2 = X_1(\lambda_3 + \lambda_4 + \ldots)$

4 Tracing the System Reliability Indices of a Power System

4.1 Tracing the Reliability Index LOLP

Based on the reliability evaluation principle, the system index LOLP for a power system is given by [2, 3, 19]

$$\mathrm{LOLP} = \sum_{S \in \Theta} P_{(S)} \tag{17}$$

where Θ is the set of all system failure states with loss of load in a given time interval.

Based on Eqs. (6) and (17), the tracing result of the reliability index LOLP for component i with all system states is given by

$$\mathrm{LOLP}_i = \sum_{S \in \Theta} P_{(S \to i)} \tag{18}$$

4.2 Tracing the Reliability Index LOLF

Similarly, the reliability index LOLF for a power system is given by [2, 3, 19]

$$\mathrm{LOLF} = \sum_{S \in \Theta} F_{(S)} \tag{19}$$

According to Eq. (13), the tracing result of the reliability index LOLF for component i with all system states is given by

$$\mathrm{LOLF}_i = \sum_{S \in \Theta} F_{(S \to i)} \tag{20}$$

4.3 Tracing the Reliability Index EENS

Similarly, the reliability index EENS for a power system is given by [2, 3, 19]

$$\mathrm{EENS} = \sum_{S \in \Theta} \mathrm{EENS}_{(S)} = \sum_{S \in \Theta} L_{(S)} P_{(S)} T, \tag{21}$$

where $\mathrm{EENS}_{(S)}$ is the EENS index of the system failure state S; $L_{(S)}$ is the load curtailment of the system failure state S, and T is the time duration of the given time interval.

The tracing result of the reliability index EENS for component i with the system state S is given by

$$\mathrm{EENS}_{(S \to i)} = L_{(S)} P_{(S \to i)} T \tag{22}$$

The tracing result of the reliability index EENS for component i with all system states is given by

$$\mathrm{EENS}_i = \sum_{S \in \Theta} \mathrm{EENS}_{(S \to i)} \tag{23}$$

4.4 UT Indices for Individual Components

The UT indices for individual components are defined as follows:

$$\mathrm{LOLP}_i\,\% = \frac{\mathrm{LOLP}_i}{\mathrm{LOLP}} \times 100\,\% \tag{24}$$

$$\mathrm{LOLF}_i\% = \frac{\mathrm{LOLF}_i}{\mathrm{LOLF}} \times 100\,\% \tag{25}$$

$$\mathrm{EENS}_i\% = \frac{\mathrm{EENS}_i}{\mathrm{EENS}} \times 100\%. \tag{26}$$

The UT indices, i.e. $\mathrm{LOLP}_i\%$, $\mathrm{LOLF}_i\%$ and $\mathrm{EENS}_i\%$, indicate the percentage contribution of component i to the system unreliability indices. In other words, a high magnitude of the $\mathrm{LOLP}_i\%$, $\mathrm{LOLF}_i\%$ and $\mathrm{EENS}_i\%$ for a component indicates high impact of the component on the overall system unreliability. The components with large $\mathrm{LOLP}_i\%$, $\mathrm{LOLF}_i\%$, and $\mathrm{EENS}_i\%$ are the WPs of a power system.

5 Algorithm for Tracing the Unreliability of a Power System

5.1 Algorithm for Tracing the Unreliability of HL-I

The proposed UT algorithm for power system at HL-I assessment can be summarized as follows:

Step 1. Enumerate a system state S based on the state of the components.
Step 2. Calculate the reliability indices of system state S, i.e., LOLP, LOLF, and EENS.
Step 3. Calculate the tracing result of the reliability index for any component i with state S, i.e., $\mathrm{LOLP}_{(S \to i)}$, $\mathrm{LOLF}_{(S \to i)}$, $\mathrm{EENS}_{(S \to i)}$.
Step 4. Repeat Step 1 to Step 3 and calculate the tracing result of the reliability index for any component i for all the states using (18), (20), (23).
Step 5. Calculate the UT indices using Eqs. (24), (25), (26) and obtain the WPs of the system.

5.2 Algorithm for Tracing the Unreliability of HL-II

There are five basic steps for adequacy evaluation of power systems at HL-II assessment:

Step 1. Select or sample a system state based on the component states.
Step 2. Analyze if the system state is a failure state, which results in load curtailment, and calculate the reliability indices associated with the failure state.
Step 3. Update cumulative bus and system indices.
Step 4. Go to Step 1 if all the system states are not considered.
Step 5. Evaluate the bus and system reliability indices, and output the results.

The impact assessment of each component on the system unreliability is embedded in the process of failure state analysis based on the above adequacy evaluation process. The contribution of each component failure to the system unreliability index is different in each failure state of a power system. The conventional algorithm for the adequacy evaluation of a power system is modified to obtain the flowchart in Fig. 2 for tracing the unreliability of a power system with N system states using the contingency enumeration approach.

It can be seen from Fig. 2 that the flowchart is similar to the traditional adequacy assessment process for a power system. The difference is that the process of tracing and cumulating the unreliability indices of system failure states are incorporated in the modified algorithm and are shown in Fig. 2 by dashed lines. The proposed UT methodology can therefore be easily implemented on existing reliability evaluation programs with only a little additional computation requirement.

6 Case Studies

6.1 Unreliability Tracing for HL-I

A computer program coded in VC++ has been developed based on the proposed algorithm. The RBTS, IEEE-RTS at HL-I are used to demonstrate the effectiveness of the proposed method for tracing the LOLP, LOLF, and EENS indices.

- RBTS Analysis

The test system developed at the University of Saskatchewan designated as the RBTS [1] was studied. The RBTS has 11 generating units and the peak load is 185 MW. The same load model is used for both the RBTS [1] and the IEEE-RTS [20]. The relevant generating unit reliability parameters are shown in Table 1.

System states with up to five components on outage were investigated. The total probability of the investigated system states is 0.9999999853. Based on the

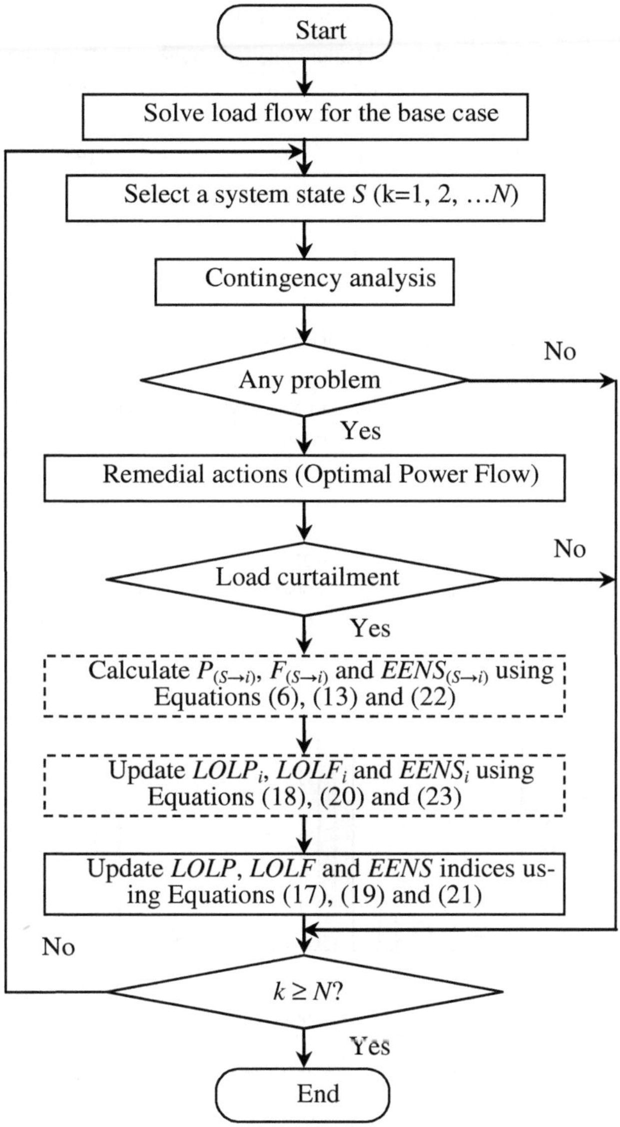

Fig. 2 Flowchart for tracing the unreliability of a power system using the contingency enumeration approach

proposed UT technique, the UT results for the RBTS annual indices are shown in Table 2 and Fig. 3.

The results show that the sum of every index for all components equals the system index. The component set with the greatest impact on the system reliability index LOLP consists of G3 (33.187 %), G4 (33.187 %), and G11 (19.479 %) based

Table 1 Generating unit reliability parameters for the RBTS

Unit No.	Cap./(MW)	Failure rate (occ./year)	Repair rate (occ./year)
G1	10	4	196
G2	20	5	195
G3, G4	40	6	194
G5, G6	5	2	198
G7, G8, G9, G10	20	2.4	157.6
G11	40	3	147

Table 2 Unreliability tracing results for the RBTS at a peak load of 185 MW

Unit No.	Tracing the unreliability indices for the units			UT indices		
	LOLE (h/year)	LOLF (occ./year)	EENS (MWh/year)	$LOLP_i$%	$LOLF_i$%	$EENS_i$%
G1	0.0152	0.00133	0.140	1.385	2.224	1.409
G2	0.0495	0.00346	0.519	4.510	5.794	5.239
G3, G4	0.3640	0.01810	3.206	33.187	30.305	32.376
G5, G6	0.0027	0.00041	0.023	0.242	0.682	0.234
G7–G10	0.0213	0.00174	0.221	1.942	2.911	2.227
G11	0.2136	0.01097	1.904	19.479	18.364	19.223
System	1.0968	0.05974	9.903	100	100	100

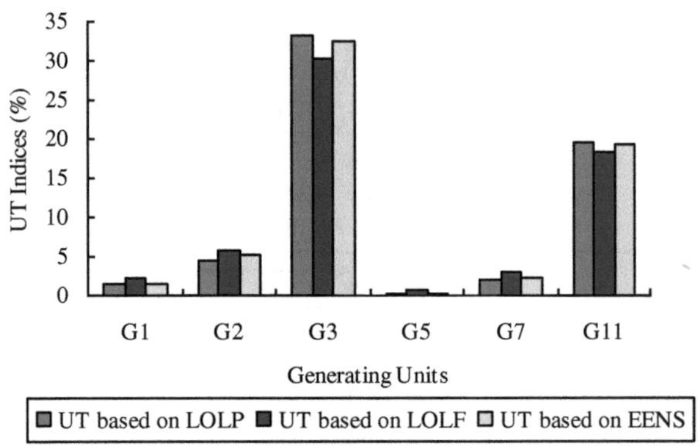

Fig. 3 UT Indices for the RBTS

on the $LOLP_i$%. $LOLF_i$%, and $EENS_i$% are analyzed in the same way. Based on the UT results and the above analysis, the WP of the RBTS HL-I are G3, G4, G11, and the unreliability (LOLP) contributions of these three components sum to 85.9 %.

In order to more clearly illustrate the reliability impact of every component on the system reliability, the peak load of the RBTS was increased to 200 and 215 MW. The UT results for the RBTS with a peak load of 200 MW are shown in Table 3. The UT result comparison between the original and modified RBTS are shown in Fig. 4.

Table 3 Unreliability tracing results for the RBTS at a peak load of 200 MW

Unit No.	Tracing the unreliability indices for the units			UT indices		
	LOLE (h/year)	LOLF (occ./year)	EENS (MWh/year)	$LOLP_i$%	$LOLF_i$%	$EENS_i$%
G1	0.0516	0.00399	0.513	1.409	2.181	1.347
G2	0.2015	0.01189	1.906	5.497	6.505	5.004
G3, G4	1.1776	0.05438	12.441	32.130	29.754	32.663
G5, G6	0.0094	0.00129	0.092	0.256	0.708	0.242
G7–G10	0.0875	0.00586	0.829	2.389	3.207	2.175
G11	0.6878	0.03210	7.289	18.767	17.564	19.138
System	3.6651	0.18277	38.089	100	100	100

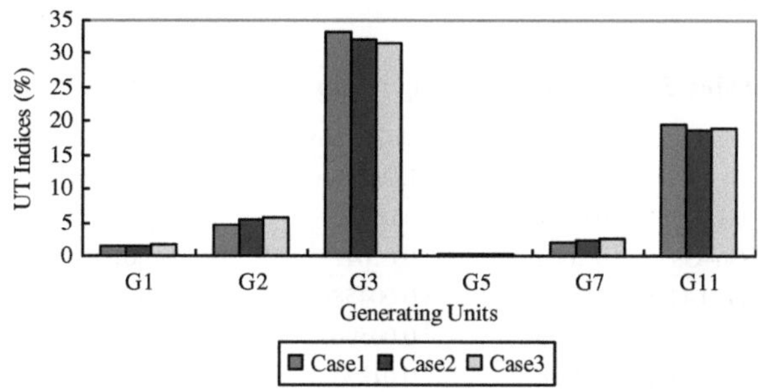

Fig. 4 UT indices based on LOLP for the RBTS. *Case 1*, *Case 2* and *Case 3* are the RBTS and modified RBTS with peak loads of 200 and 215 MW respectively

Though the contribution to the LOLP, LOLF, and EENS of the large rated capacity units at a load of 200 MW are larger than those at 185 WM, the $LOLP_i$%, $LOLF_i$%, and $EENS_i$% values are lower than those at 185 MW. In other words, the relative reliability impact of the large rated capacity units reduces as the peak load increases.

Three different contingencies levels ranging from 4 to 6 were utilized to investigate the effect of the contingency level on the UT. The comparison results are presented in Fig. 5. The LOLE values based on 4, 5, and 6 contingencies are 1.0938, 1.0968 and 1.0969 h/year, respectively. It can be seen from Fig. 5 that increasing the contingency level has only a small effect on the UT indices for the RBTS.

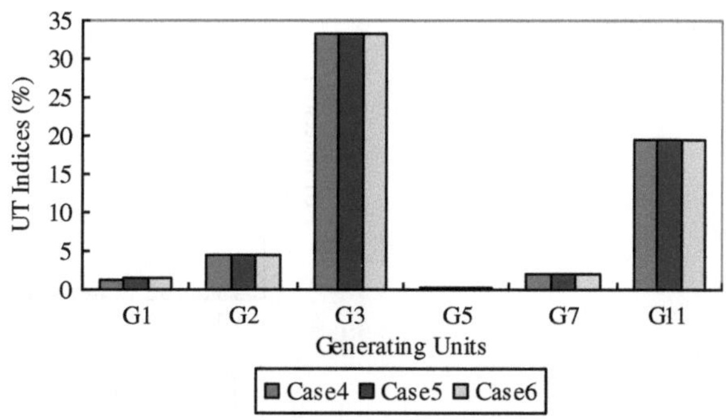

Fig. 5 UT indices based on LOLP for RBTS. *Case 4*, *Case 5* and *Case 6* are the RBTS values with contingencies level 4, 5 and 6 respectively

Table 4 Generating unit reliability parameters for the IEEE-RTS

Unit No.	Cap./(MW)	MTTF (h/occ.)	MTTR (h/occ.)
G1–G5	12	2940	60
G6–G9	20	450	50
G10–G15	50	1980	20
G16–G19	76	1960	40
G20–G22	100	1200	50
G23–G26	155	960	40
G27–G29	197	950	50
G30	350	1150	100
G31–G32	400	1100	150

Table 5 Unreliability tracing results for the IEEE-RTS

Unit No.	Tracing the unreliability indices for the units			UT indices		
	LOLE (h/year)	LOLF (occ./year)	EENS (MWh/year)	$LOLP_i$%	$LOLF_i$%	$EENS_i$%
G1–G5	0.0094	0.00242	0.962	0.113	0.367	0.098
G6–G9	0.2044	0.02047	21.601	2.441	3.103	2.196
G10–G15	0.0033	0.00458	0.355	0.04	0.695	0.036
G16–G19	0.0159	0.00596	1.742	0.19	0.903	0.177
G20–G22	0.0724	0.01275	8.049	0.864	1.932	0.818
G23–G26	0.1103	0.02312	12.673	1.317	3.503	1.288
G27–G29	0.2215	0.03198	26.175	2.645	4.847	2.661
G30	1.1626	0.07341	143.28	13.885	11.126	14.564
G31, G32	2.4698	0.10723	293.409	29.497	16.251	29.825
System	8.3731	0.65983	983.770	100	100	100

- IEEE-RTS Analysis

The IEEE-RTS was developed by an IEEE Task Force to provide a practical representative BPS for research and comparative study purposes. The IEEE-RTS has 32 generating units and the peak load is 2850 MW. The relevant reliability parameters are shown in Table 4 [20]. System states with up to five components on outage were investigated. The UT results are shown in Table 5 and Fig. 6.

Fig. 6 UT Indices for the IEEE-RTS

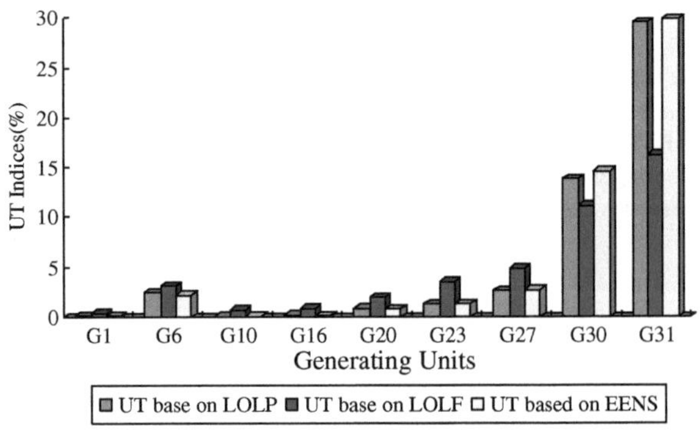

Table 5 shows that the components with the greatest impact on the system reliability index LOLP are G31 (29.497 %), G32 (29.497 %), G30 (13.885 %) based on the $LOLP_i$%. The WPs of the IEEE-RTS at HL-I are therefore the G31, G32, and G30.

The UT result for G31 $LOLP_i\% = 29.497\ \%$ is larger than that for $LOLF_i\% = 16.251\ \%$. The contribution of G31 to the LOLF index is therefore smaller than that to the LOLP index. This is due to the fact that G31 while a relatively large unreliability has a relatively small repair rate.

6.2 Unreliability Tracing for HL-II

A computer program coded in C++ based on the algorithm shown in Fig. 2 was developed to trace the component unreliability contributions of a BPS. A DC flow-based optimal load shedding approach [2, 3, 19] is used to determine the load curtailments in reliability evaluation process using the contingency enumeration approach. The RBTS and the IEEE-RTS are used to test and validate the UT technique for BPS.

The single line diagram of the RBTS is shown in Fig. 1. The RBTS has 11 generating units, 9 transmission lines, and a peak load of 185 MW. The detailed system data are given in Reference [1]. In the analysis conducted for the RBTS, independent overlapping outages up to the fifth level for generating units and up to the fourth level for transmission lines or composite components are considered. A 20-step monotone increasing model of the load duration curve using the IEEE-RTS annual chronological hourly load profile was used for the calculation of annual system indices [1, 2].

The IEEE-RTS has 32 generating units, 33 transmission lines, 5 transformers, and a peak load of 2850 MW. The detailed system data and single line diagram are given in Reference [20]. In the analysis conducted for the IEEE-RTS, independent overlapping outages up to the fourth level for generating units and up to the third level for transmission lines or composite components are considered.

The annualized system indices of the RBTS and the IEEE-RTS, such as LOLP, LOLF and EENS, are shown in Table 6. The annual system indices of the RBTS at different peak load levels are also shown in Table 6.

- UT Analysis of Annualized System Indices for the RBTS

The UT results of annualized system indices for the RBTS generation and transmission line components are shown in Table 7 and Table 8, respectively. It can be seen from Table 7 and Table 8 that the sum of all component UT indices, such as $LOLP_i$%, $LOLF_i$% and $EENS_i$%, equals to 100 %. In other words, the proposed UT technique can completely distribute the reliability indices of a BPS to the components.

Table 6 Annualized and annual system indices for the RBTS and IEEE-RTS

–	Systems	Peak load (MW)	LOLP	LOLF (Occ./Year)	EENS (MWh/Year)
Annualized system indices	The RBTS	185	0.00976	5.51421	1052.36
	The IEEE-RTS	2850	0.07527	51.57058	105057.76
Annual system indices	The RBTS	185	0.00127	1.32540	134.55
	The RBTS	210	0.00219	1.74818	241.55
	The RBTS	160	0.00115	1.26365	107.30

Table 7 Generating unit UT results of the annualized system indices for the RBTS

Generating No.	Bus	Capacity (MW)	$LOLP_i$	$LOLF_i$ (Occ./Year)	$EENS_i$ (MWh/Year)	$LOLP_i$%	$LOLF_i$%	$EENS_i$%
G1	1	40	0.00230	0.86347	265.17	23.540	15.659	25.198
G2	1	40	0.00230	0.86347	265.17	23.540	15.659	25.198
G3	1	10	0.00013	0.16165	18.75	1.314	2.932	1.782
G4	1	20	0.00099	0.50581	65.01	10.125	9.173	6.177
G5	2	5	0.00004	0.11913	5.15	0.446	2.160	0.489
G6	2	5	0.00004	0.11913	5.15	0.446	2.160	0.489
G7	2	20	0.00043	0.25321	27.65	4.422	4.592	2.628
G8	2	20	0.00043	0.25321	27.65	4.422	4.592	2.628
G9	2	20	0.00043	0.25321	27.65	4.422	4.592	2.628
G10	2	20	0.00043	0.25321	27.65	4.422	4.592	2.628
G11	2	40	0.00126	0.43869	147.66	12.935	7.956	14.031
Generation system			0.00879	4.08419	882.66	90.034	74.067	83.875

It can also be seen that the same generating units installed at the same bus, such as G1 and G2, have a same UT results. This is because they have the same locations, reliability, and electric parameters. Therefore, they have the same contributions to the system unreliability. The parallel transmission lines, such as L1 and L6 or L2 and L7, have the similar UT results. Tables also show that the generation system amounts 90.0, 74.1, and 83.9 % for the annualized LOLP, LOLF, and EENS indices, respectively. The unreliability of the RBTS, is therefore, mainly dominated by the generating units.

The unreliability contributions of G1, G2, G11, and L9 for the annualized LOLP and EENS indices sum to 69.4 and 79.6 %, respectively. Therefore, the four components are the WPs of the RBTS at a peak load of 185 MW. This is due to the fact that G1, G2, and G11 are the largest rating capacity units with the highest unavailability parameters, and L9 is the only component that does not meet the N-1 criterion.

Table 8 Transmission line UT results of the annualized system indices for the RBTS

Branch No.	From Bus	To bus	$LOLP_i$	$LOLF_i$ (Occ./Year)	$EENS_i$ (MWh/Year)	$LOLP_i$%	$LOLF_i$%	$EENS_i$%
L1	1	3	0.00001	0.12583	1.11	0.073	2.282	0.105
L2	2	4	0.00002	0.06615	2.95	0.188	1.200	0.280
L3	1	2	0.00000	0.03985	0.75	0.047	0.723	0.071
L4	3	4	0.00000	0.00834	0.10	0.006	0.151	0.009
L5	3	5	0.00000	0.00938	0.28	0.011	0.170	0.027
L6	1	3	0.00001	0.12583	1.11	0.073	2.282	0.105
L7	2	4	0.00002	0.06615	2.95	0.188	1.200	0.280
L8	4	5	0.00000	0.00940	0.28	0.011	0.171	0.027
L9	5	6	0.00091	0.97908	160.18	9.367	17.756	15.221
Transmission system			0.00097	1.43002	169.7	9.966	25.933	16.125

Note "0.00000" indicates that the value is less than 1E-5

- UT Analysis of Annual System Indices for the RBTS

The UT results of the annual system indices for the RBTS at a peak load of 185 MW are shown in Table 9. In this case, the generating system has a relatively small "contribution" to the system unreliability compared with the annualized results in Table 7. The annualized results are obtained using only the system peak load as the load model, and therefore, the generation system is inadequate to continuously satisfy the system peak load. The lower load levels are also considered in the annual results during which period the generation is adequate, and therefore, their contribution to the system unreliability is greatly reduced.

The UT results for the annual system indices of the RBTS at the peak load of 210 and 160 MW are shown in Table 10. It can be seen from Tables 6, 9 and 10 that the reliability indices and UT indices change significantly with the peak load.

Table 9 The UT results of the annual system indices for the RBTS at a peak load of 185 MW

Components	$LOLP_i$%	$LOLF_i$%	$EENS_i$%
G1, G2	7.293	4.285	6.963
G3	2.804	3.155	2.769
G4	2.990	3.053	2.995
G5, G6	0.558	1.699	0.552
G7, G8, G9, G10	0.676	1.093	0.668
G11	2.113	1.245	1.916
L1, L6	0.084	0.228	0.083
L2, L7	0.371	0.521	0.374
L3	0.272	0.394	0.275
L4	0.042	0.092	0.042
L5	0.083	0.175	0.126
L8	0.083	0.176	0.126
L9	72.298	73.870	73.135

Table 10 UT results of the annual system indices for the RBTS at different peak load levels

Peak loads	210 MW			160 MW		
Components	$LOLP_i$%	$LOLF_i$%	$EENS_i$%	$LOLP_i$%	$LOLF_i$%	$EENS_i$%
G1, G2	16.997	8.420	15.533	4.865	3.290	4.847
G3	2.470	3.303	2.466	2.864	3.120	2.863
G4	3.867	3.842	3.983	2.824	2.912	2.812
G5, G6	0.506	1.799	0.517	0.566	1.683	0.566
G7, G8, G9, G10	1.333	1.627	1.350	0.540	0.986	0.534
G11	8.839	3.728	7.705	0.495	0.588	0.486
L1, L6	0.911	1.814	0.551	0.083	0.139	0.084
L2, L7	0.298	0.781	0.316	0.388	0.461	0.391
L3	0.166	0.508	0.181	0.295	0.365	0.296
L4	0.026	0.119	0.028	0.046	0.089	0.046
L5	0.049	0.178	0.080	0.091	0.177	0.137
L8	0.050	0.182	0.081	0.091	0.177	0.137
L9	41.777	56.005	46.244	79.328	77.48	79.314

The UT indices of the transmission lines increase significantly, whereas the UT indices of the generating units decrease, with the decrease in the system peak loads. Since there is adequate generation at a relatively low peak load, such as 160 MW, the UT indices of transmission lines are higher than those at a relatively high peak load. This conclusion is drawn from the RBTS studies. However, the results can change if the increase in load creates transmission congestions. The results are system dependent, and the developed methodology can be applied to any system.

It should be noted that the transmission line L9 is the WP of the RBTS irrespective of the change in the peak load. This is due to the fact that the failure of L9 always results in the loss of the load at Bus 6 at any peak load level. Adding a line parallel to L9 should be considered if a reliability upgrade project is planned for the system. With this upgrade, the annual system indices of the modified RBTS, i.e. the LOLP, LOLF and EENS, change to 0.00013, 0.07482 and 11.98, respectively. In other words, improving the reliability performance of L9 can provide significant improvement in the overall system reliability. The WP of the modified RBTS is no longer L9 as in the original RBTS, but is identified as G1 and G2.

- UT Analysis of Annualized System Indices for the IEEE-RTS

The UT results of the annualized system indices for the IEEE-RTS are shown in Table 11. For simplicity, Table 11 only presents the UT of individual generating units as the contributions of the transmission lines and transformers to the system unreliability are relatively small.

Table 11 shows that the unreliability of the IEEE-RTS is also largely dominated by the generating units, and this is consistent with the results in Reference [6]. The WPs of the IEEE-RTS are G22 and G23, which amount 53.8 and 59.1 % for the annualized LOLP and EENS indices, respectively. This is largely due to the fact

Table 11 UT results of the annualized system indices for the IEEE-RTS

Generating units	Bus	Capacity (MW)	$LOLP_i$	$LOLF_i$ (Occ./Year)	$EENS_i$ (MWh/Year)	$LOLP_i$%	$LOLF_i$%	$EENS_i$%
G1, G2	1	20	0.00254	2.01049	2657.51	3.373	3.899	2.530
G3, G4	1	76	0.00018	0.52714	214.40	0.243	1.022	0.204
G5, G6	2	76	0.00018	0.52714	214.40	0.243	1.022	0.204
G7, G8	2	20	0.00254	2.01045	2657.51	3.373	3.898	2.530
G9–G11	7	100	0.00070	0.93473	910.99	0.933	1.813	0.867
G12–G14	13	197	0.00281	2.80779	2867.81	3.736	5.445	2.730
G15–G19	15	12	0.00014	0.29116	131.50	0.184	0.565	0.125
G20	15	155	0.00109	1.68832	1300.06	1.447	3.274	1.237
G21	16	155	0.00109	1.68840	1300.06	1.447	3.274	1.237
G22	18	400	0.02024	7.30272	31033.98	26.893	14.161	29.540
G23	21	400	0.02023	7.29652	31031.98	26.879	14.149	29.538
G24–G29	22	50	0.00004	0.51383	47.88	0.055	0.996	0.046
G30, G31	23	155	0.00109	1.68835	1300.06	1.447	3.274	1.237
G32	23	350	0.00793	3.91145	14018.54	10.534	7.585	13.344
Generation system			0.07513	51.1809	105053.55	99.814	99.244	99.996
Transmission system			0.00014	0.38967	4.21	0.186	0.756	0.004

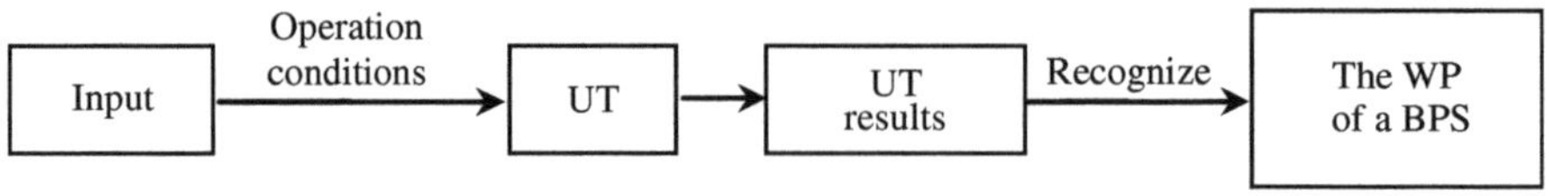

Fig. 7 Identifying the WP of BPS using the UT technique

that G22 and G23 are the largest rating capacity generating units with the highest unavailability. Reducing the failure rate or repair time of G22 and G23 is an effective measure to improve the reliability performance of the IEEE-RTS.

The contribution of G22 and G23 to the LOLF index (14.161 and 14.149 %) is smaller than that to the LOLP (26.893 and 26.879 %) and the EENS (29.540 and 29.538 %) indices. This is because these two generating units have a relatively large unreliability and a relatively small repair rate.

The proposed technique has been tested on other practical BPSs, such as the Guangdong BPS in China with more than 600 buses and 1500 components. The UT results indicate that more than 90 % of the system unreliability is contributed from less than 10 % of the system components that are identified as the WP of Guangdong BPS. The BPS planners and operators can therefore use the WP information to focus on the WP to improve the system reliability and alleviate the system risk.

It is well known that the system reliability usually varies with the operating conditions, and therefore, the UT indices and the WP of a BPS also vary with the operation conditions. Using a step-by-step procedure shown in Fig. 7, the proposed UT technique for recognizing the WP can be performed to identify the WP of a BPS considering the operating conditions relevant to any time period of interest. The valuable WP information can be obtained from different UT results of various operating conditions.

7 Conclusions

Two unreliability tracing principles, i.e. failure component sharing principle (FCSP) and proportional sharing principle (PSP), are presented to derive the UT model and algorithm for a power system. The proposed technique provides a new idea for evaluating the impact of components on the system unreliability and can be used to distribute the unreliability indices of a power system to the system components impartially and reasonably. The models and algorithms for tracing the unreliability are determined using the PSP. The UT indices for individual component are created to recognize the major unreliability contributions (MUC) of the power system. The method provides the system planner/operator with information on how much an individual component contributes to the overall system risk and the system MUC.

The basic principles and methods proposed in this paper are general and therefore applicable to both HL-I and HL-II. The illustration application is at HL-I and HL-II, and the RBTS and IEEE-RTS were examined using the proposed method. The results show that the proposed approach has good practicability and has promising prospects in other engineering applications.

Power system applications utilizing the proposed UT technique can be used to recognize weak system elements, improve the reliability, analyze reliability cost/benefits and optimize scheduled maintenance. In addition, the proposed UT technique can be easily implemented on the existing reliability evaluation programs with a relatively little additional computation requirement, and conveniently expanded to AC/DC systems, distribution systems, or interconnected systems.

References

1. Billinton R et al (1989) A reliability test system for educational purposes—Basic data. IEEE Trans Power Syst 4(3):1238–1244
2. Billinton R, Li W (1994) Reliability assessment of electric power systems using Monte Carlo methods. Plenum Press, New York
3. Li W (2004) Risk assessment of power systems: models, methods and applications, IEEE Press and John Wiley & Sons Inc
4. Billinton R, Goel L (1992) Overall adequacy assessment of an electric power system. IEE Proc Gener Transm Distrib 139(1):57–63
5. Goel L, Billinton R (1995) Pertinent factors in the adequacy evaluation of an overall electric power system. IEE Proc Gener Trans Distrib 142(4):337–342
6. Pinheiro JMS, Dornellas CRR et al (1998) Probing the new IEEE reliability test system (RTS-96): HL-II assessment. IEEE Trans Power Syst 13(1):171–176
7. Billinton R, Wang P (1999) Deregulated power system planning using a reliability network equivalent technique. IEE Proc Gener Trans Distrib 146(1):25–30
8. Chowdhury AA, Koval DO (2004) Considerations of relevant factors in setting distribution system reliability standards. IEEE PES Meet 1–7
9. Hur D, Yoo C-I, Kim BH et al (2004) Transmission embedded cost allocation methodology with consideration of system reliability. IEE Proc Gener Trans Distrib 51(4):427–432
10. Chung K-H, Kim BH, Hur D et al (2005) Transmission reliability cost allocation method based on market participants reliability contribution factors. Electr Power Syst Res 73(1): 31–36
11. Melo ACG, Pereira MVF (1995) Sensitivity analysis of reliability indices with respect to equipment failure and repair rates. IEEE Trans Power Syst 10(2):1014–1021
12. Oliveira AM, Melo ACG, Pinto LMVG (1999) Consideration of equipment failure parameter uncertainties in bus composite reliability indices. In: Proceedings of IEEE power engineering society winter meeting, vol 1, New York, USA, pp 448–453
13. Billinton R, Mo R (2003) Impact of equipment availability on composite system reliability. In: Proceedings Canadian conference on electrical and computer engineering, vol 1, Montreal, Canada, pp 607–612
14. Billinton R, Nighot R (2005) Incorporating station-related outages in composite system reliability analysis. IEE Proc Gener Transm Distrib 152(2):227–232
15. Xie K, Zhou J, Billinton R (2008) Tracing the unreliability contributions of power system components. Electric Power Compon Syst 36(12):1299–1309
16. Xie K, Billinton R (2009) Tracing the unreliability and recognizing the major unreliability contribution of network components. Reliab Eng Syst Saf 94(5):927–931
17. Xie K, Hu B, Karki R (2011) Tracing the component unreliability contributions and recognizing the weak parts of a bulk power system. Eur Trans Electr Power 21(1):254–262

18. Bialek J (1996) Tracing the flow of electricity. IEE Proc Gener Trans Distrib 143(4):313–320
19. Billinton R, Allan RN (1996) Reliability of electric power system. Plenum Press, New York, USA
20. Reliability Test System Task Force of the Application of Probability Methods Subcommittee (1979) IEEE reliability test system. IEEE Trans Power App Syst PAS-98(6):2047–2054

Chapter 6
Sustainable Energy Optimization in a Smart Microgrid

Ryan Jansen and Rajesh Karki

Abstract This chapter titled “Sustainable Energy Optimization in a Smart Microgrid” presented the methodology for developing a smart microgrid optimization model that can be used to determine the most financially economical combination of microgrid technologies with acceptable level of system reliability. The smart microgrid optimization model assessed various combinations of PV and wind turbine renewable generation in increments of 5 kW varying between 0 and 200 kW. Battery energy storage was also considered in combination with renewable generation in increments of 5 kWh from 0 to 200 kWh. The smart microgrid optimization model was developed in the MATLAB environment and incorporated a smart microgrid management system to achieve additional fuel savings and increased system reliability by utilizing load shedding and load deferral techniques. Microgrid power system reliability was assessed using an SMCS simulation model. The accuracy of the model was verified using the SIPSREL program developed at the University of Saskatchewan. Results indicated that the implementation of renewable generation, battery energy storage and DSM techniques can substantially reduce the lifetime operational costs of a microgrid while increasing power system reliability. Implementation of energy storage technologies to reduce fossil fuel consumption as well as to increase renewable generation penetration is an area of ongoing research, as is the application of microgrid technology to improve remote power system reliability.

1 Introduction

A microgrid is a set of electrical power generation sources that are networked together to meet the energy needs of a localized community, but may also maintain a single connection point to a larger electrical grid [1]. Microgrids are typically

R. Jansen (✉) · R. Karki
Department of Electrical and Computer Engineering, University of Saskatchewan, Saskatoon, Canada
e-mail: Ryan.Jansen@src.sk.ca

N.R. Karki et al. (eds.), *Sustainable Power Systems*, Reliable and Sustainable Electric Power and Energy Systems Management,
DOI 10.1007/978-981-10-2230-2_6

large institutions such as prisons, hospitals, universities, etc., but they can also be small communities, or even single residence dwellings [2].

Microgrids characteristically have a high level of reliability when connected to a macrogrid and contribute to a greater reliability of the macrogrid as a whole [3] if they have been designed to support the electrical infrastructure outside of the microgrid. A microgrid requires some level of power management when operating independently to ensure that its power supply can meet all of the internal load demands, or at least some critically identified loads. The management of energy generation and usage applying automation driven by relevant data acquired and processed using digital devices is often referred to as a "smart grid" application.

Microgrids are becoming increasingly popular as a form of distributed power generation typically employing renewable energy [4]. This is largely due to the improving economic viability of renewable energy sources [5], the interest in improved sustainability and the desire to increase distributed generation to reduce loading on over-capacity distribution lines [4].

Microgrids present unique niche opportunities to incorporate renewable energy. Often, the peak load of a microgrid is quite limited depending on the size of the system. When considering a small peak load (such as below 100 kW), the economics are rather different than that of a large utility grid. On a large scale, such as a large utility grid, it is difficult for renewable energy with incentives to be cost-competitive with natural gas and coal generation stations, especially once reliability, lifetime and capacity factor are considered. For instance, Ref. [6] indicates that the levelized cost of wind generation is 29 % greater than conventional combined cycle natural gas generation.

In addition, the natural gas-fired generation is dispatchable and has an 87 % capacity factor compared to a typical capacity factor of wind being 34 % [7]. In contrast, the small-scale microgrid economics for renewable energy integration can be quite different.

Reference [8] explains that the reasons for developing small-scale microgrids are predominantly financially motivated whether it is to offset fossil fuel consumption that must be delivered to remote locations by ice road or airplane, or to reduce reliance upon a utility-power grid with prohibitively high electricity rates. Reference [9] found that microgrids consisting only of diesel generators were cost-competitive with renewable generation when fuel prices were below $0.60/L, however, renewables, such as solar, became more economical when fuel prices exceeded $0.80/L. Solar penetration, or the ratio of installed solar generation capacity to the total installed generation power capacity of the system, as high as 21 % was found to be economically beneficial in the study. Although higher degrees of penetration were tested, a growing amount of unutilized energy rendered the system uneconomical. As such, there exists a diesel-renewable balance for a small microgrid where the lifetime cost of the system is lower than alternative options including a fully renewable microgrid, and a stand-along diesel-based microgrid as described in Ref. [10]. Reference [11] found that depending on the cost of fuel, energy storage and wind generators, a viable business case can be developed when annual wind speeds are in excess of 6.0 m/s for wind generators only, and wind generators with energy storage become viable as annual wind speeds approach 7.0 m/s as long as the cost of energy storage systems is less than $1,000/kW. In the case of wind energy storage systems, the wind turbine array could be sized to produce an additional 50 % of energy for the microgrid. Reference [12] explored the reliability detriments of meeting additional system load by installing renewable generation rather than traditional diesel units in an off-grid application. It was determined that the most cost-effective method of expanding a microgrid's generation portfolio was by adding wind-diesel systems. The wind generators were used to minimize diesel consumption, while the diesel generators provided the microgrid with the desired level of reliability and the ability to meet the microgrid's stability criteria. There exist other possible methods of reducing the lifetime cost of microgrids as well. Reference [13] discussed using demand side management (DSM) techniques to modify the energy consumption of electric customers in order to smooth out the daily load peaks and valleys. The resultant calculations determined that using DSM techniques, the generation components for a small off-grid community could be reduced by up to up to 20 %, and the addition of photovoltaic energy sources could be used to optimize the size of a diesel generator thereby reducing the cost of the entire test system for a six-home microgrid.

Although it is clear that combinations of renewables, storage and generation can be more financially viable than any of the individual technologies on their own, there is uncertainty regarding the ideal cost-effective combination of technologies. As such, there is a need to develop appropriate models that balance all the applicable variables in determining the most beneficial combination of microgrid technologies, given available renewable resources, loads and methods of load management and energy storage.

This chapter presents the methodology and development of a smart microgrid optimization model that can be used to assess the most cost-effective combination and sizing of microgrid technologies, including conventional fossil fuel generation, renewables including wind turbines and photovoltaics, energy storage and active management of loads.

The structure of this chapter is as follows: the characteristics of microgrid technologies are presented in Sect. 2. Section 3 introduces the development of the optimization model, indicates the methodology that will be used for the management of the microgrid, and presents preliminary results without the application of a smart microgrid controller or implementation of DSM. Section 4 presents the premise for the operation of a smart microgrid controller, identifies loads that can be actively managed and explains the integration of the smart microgrid controller into the microgrid optimization model to create a smart microgrid optimization model. Results of the smart microgrid optimization model are presented in Sect. 5. A method of reliability assessment of the microgrid is introduced in Sect. 6, and conclusions are made in Sect. 7.

2 Microgrid Characteristics for Optimization Model Development

The microgrid analyzed in this project consisted of five homes with an integrated combination of electric energy from wind turbine generation, photovoltaic generation, propane generation and battery energy storage. The load characterization of the five-home system and each of the potential electrical generation and storage components are discussed in the following sections.

2.1 Load Characterization

Real-time residential load data obtained from Check-It Solutions were used in this study to obtain realistic load characteristics for typical residential customers. Check-It Solutions is a company founded in Regina, Canada that has developed an affordable, flexible and customizable method of measuring energy consumption in residential homes. Access to real-time and historical electric energy consumption was provided for five typical Saskatchewan homes to facilitate the integration of these load profiles into the algorithms that have been developed.

The data acquisition system installed by Check-It Solutions [14] provided historical energy consumption in hourly increments. A computer program was written to distribute the hourly data evenly into 15 min intervals to mirror the time increment that was used in the analysis of the power generation from the renewable sources and the energy storage facility in the system.

The overall purpose of obtaining several time-varying load profiles for a group of residences was to analyze how the cumulative time-of-day power demand for the system would vary with the addition of multiple residences. A simple stepped load profile estimated for a single home and multiplied to form a 'group' of homes would not be representative of the varying time-of-day demands of multiple unique homes. This is consistent with diversified load analysis in electrical utilities and is well understood. Figure 1 indicates the cumulative, diversified load profile of the five-home system.

The average annual energy consumption of the proposed five-home system is 13,807 kWh per home. In comparison, an energy-efficient home located in Saskatchewan has been monitored and found to consume an average of 14,000 kWh per year. This home uses electricity as its sole utility source, and has its own well water supply, and waste management system. It was therefore assumed that the acquired load profiles from Check-It Solutions were indicative of a similarly designed home with similar characteristics.

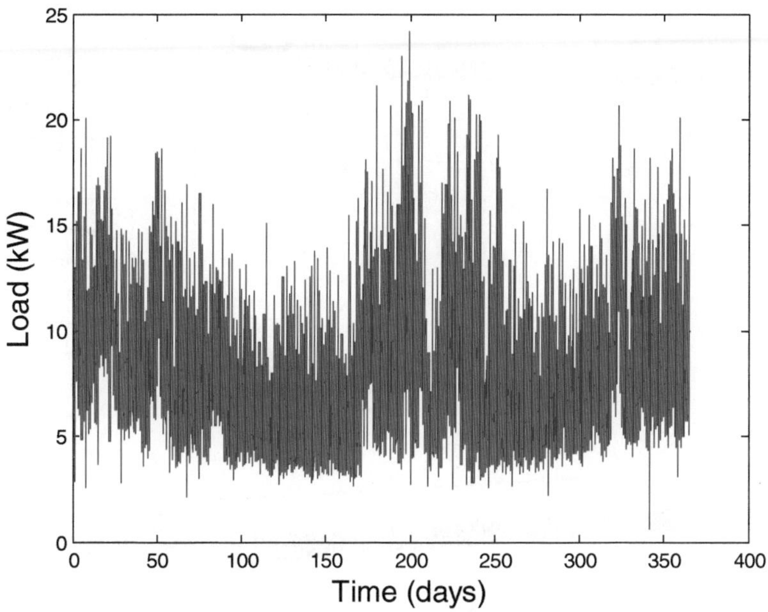

Fig. 1 Cumulative annual load profile of a five-home system

2.2 Generation and Storage Characteristics

The following portion of the chapter presents the various energy generation sources that will be considered by the developed smart grid optimization model, complete with costing and production information.

Fossil fuel generation is commonly used by microgrid systems to provide energy to a microgrid when the residential load is greater than the system's ability to provide renewable energy or electrical storage and is disconnected from the utility grid.

There are several different types of generators that operate using natural gas, propane, or diesel fuel. Propane is a common fuel used for heating in rural areas where natural gas may be too expensive to bring in. As of 2015, the cost of liquid propane was approximately \$0.72/L delivered to site in Saskatchewan.

For this study, it was assumed that a generator similar to a Generac generator was used for producing power to the microgrid system. Generac offers residential, commercial and industrial generators which range in size from 6 to 300 kW and operate on natural gas, propane or diesel. An image of a typical Generac generator is portrayed in Fig. 2.

A quotation was supplied by a local company for the procurement and installation of a 17 kW generator with the capacity to use either natural gas or liquid propane as its fuel source. The total cost of the quotation was \$8,946.15 which is equivalent to approximately \$520/kW.

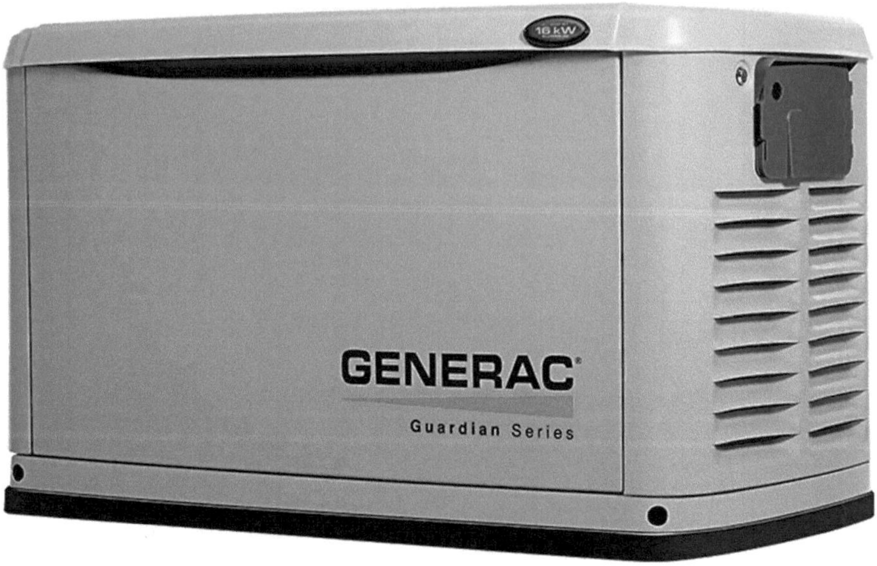

Fig. 2 Generac propane generator [15]

The liquid propane consumption of this generator is 9.73 L/h under full load [16], 6.10 L/h at 50 % load, and it will be assumed that the relationship between the load and fuel consumption is linearly dependent. This yields a fuel consumption of 0.57 L/kWh, or an electrical cost of $0.41/kWh assuming the generator operates at full capacity. In comparison, the fuel consumption at 50 % load is 0.72 L/kWh, and the electrical cost would be $0.52/kWh, an operational cost increase of 26 %.

Photovoltaic (PV) panels have decreased in price by approximately 70 % in the past ten years due to an increase in manufacturing efficiencies, consumer demand and increased collector electrical conversion efficiency. As such, many small-scale renewable energy installations include photovoltaic arrays.

A quotation for the supply and installation of a 10.42 kW array was provided by a local company that specializes in the distribution and installation of solar panels. The total cost of the system was $42,450. This included equipment, electrical installation, installation labour and foundation work. This was an equivalent cost of $4,170/kW, and yielded an energy cost of $0.16/kWh over a 20 year lifetime based on average power production. The solar panels carry a 25 year 80 % output warranty (Fig. 3).

The local company also provided an annual energy profile for the electrical production of an 8.19 kW PV array. The total annual production of the array was 10,690 kWh, and the capacity factor was 14.9 %. The annual energy production of the array was recorded using a Bluetooth data logger at 10 min intervals. A computer program was written to aggregate this energy production over a 30 min

Fig. 3 Saskatoon photovoltaic array—8.19 kW [17]

Fig. 4 Raum 3.5 kW wind turbine [18]

period, and divide it evenly into two 15 min periods to mirror the time step that was used to analyze the other forms of energy generation.

Wind power generation has become an increasingly common option for rural residents to generate their own power over the past 15 years, and will also be included in this analysis. Turbines, such as the one shown in Fig. 4, are indicative of typical small-scale wind turbines that have been installed in rural areas.

Although an annual generation profile for the 3.5 kW wind turbine was not currently available, the Saskatchewan Research Council (SRC) has done extensive monitoring of wind resources at numerous sites across Saskatchewan. SRC is a leading research, development and demonstration organization with key focuses in energy, and mining and minerals. A full year of wind speed data collected every minute was available from an installed site near Dalmeny, Saskatchewan. This 1 min data was sampled every 15 min so as to mirror the time increment that was used in the analysis by the other forms of generation and storage in the system. The resultant wind profile was applied to the measured power curve supplied by Raum [18] for the 3.5 kW wind turbine generator.

The mean annual wind speed measured at a site near Dalmeny in 2010 was 4.01 m/s and is comparable to the previous 5 year average of 4.36 m/s based on Environment Canada data from the Saskatoon airport. Using this data, the annual electrical generation for a Raum 3.5 kW wind turbine was calculated to be 3,030 kWh, and the capacity factor was 13.8 %.

The cost of the 3.5 kW Raum turbine was calculated using a quotation received for the purchase of equipment, installation and foundation work. The total cost of the system was $22,000. This is an equivalent cost of $6,280/kW and will be used in estimating the lifetime cost of the microgrid systems. The turbine carries a 5 year limited warranty, and an estimated 20 year lifetime. The equivalent energy cost based on this lifetime is calculated to be $0.36/kWh.

There are many different battery technologies that are readily accessible for mass energy storage. The most predominant of these batteries has been the lead acid battery. It is robust, inexpensive and well proven; however, its limited lifetime and depth of discharge is a deterrent to its use in this project as the battery may be required to cycle up to two to three times per day. In light of these requirements, the next most well-proven battery type is the lithium-ion battery.

Lithium-ion batteries have recently decreased in price while their characteristics have been improving. Companies have published that their batteries are capable of enduring >8,000 cycles at 80 % depth of discharge [19], and this can be extended to 90 % depth of discharge. Also integral to the lithium-ion battery system, a programmable logic controller monitors the state of charge, state of health, rate of charge and discharge, internal temperature and individual cell energy balance of the batteries to optimize their performance.

Prices for lithium ion batteries vary greatly and depend on the application, size and requirement for power electronics such as inverters. Costing information gathered for this project showed cost variations ranging from $420 to $2,000/kWh. An article [20] stated that lithium ion batteries were generally found to be between $500 and $1,000/kWh, and were forecasted to decrease 50–75 % over the next 5–10 years.

For the purposes of this project, it will be assumed that a lithium ion battery complete with PLC-based inverter will cost $750/kWh. The battery is assumed to have the capability to achieve 90 % depth of discharge, and have a round trip efficiency of 90 %.

3 Optimization Model Development and Base Case Results

This section discusses the development of the optimization model in the MATLAB [21] environment and presents a cost-comparison for different combinations of renewable energy, battery energy storage and fossil fuel generation.

As presented, studies show that renewable generation is capable of producing electricity at a cost less than conventional generators at a microgrid-scale. As well, there is opportunity to size the renewable energy capacity such that excess energy generated at the site can be stored via energy storage technologies, thereby further decreasing the fuel consumed by onsite fossil fuel generators.

The methodology of the optimization model is as follows. During periods when renewable power is available, it will serve the load. If the renewable generation is not sufficient to meet the microgrid load, battery energy will be dispatched as needed down to 10 % state of charge (SOC) up to a rate of 25 kW. If the available renewable generation is greater than the microgrid load, the battery will be charged up to 100 % SOC at a maximum rate up to 25 kW. Finally, if the renewable generation and the available battery capacity is not sufficient to meet the microgrid load, the propane generator will ramp up accordingly to meet the generation requirement. The round trip efficiency of the battery is considered to be 90 %.

The optimization model analyzed a number of combinations of battery energy storage, wind turbine power and PV generation capacities. Specifically, it varied combinations of wind turbine and PV generation by 5 kW increments from 0 to 200 kW, and of battery energy capacity between 0 and 200 kWh in 5 kWh increments. With each combination, a 20 year lifetime cost was calculated based on the specified renewable generation size, battery energy storage and generator size and fuel consumption. An economically optimal base case was first analyzed without incorporating intelligent control of the microgrid.

Figure 5 describes the conceptual layout and the control method of the base case microgrid. It should be noted that a transfer switch is present in Fig. 5 as a potential connection to a macrogrid. Although this is common feature in microgrids, the subsequent analysis for the optimized base case and the ideal microgrid with smart microgrid controller does not consider that a macrogrid connection is available. For the purposes of this research, the microgrid community is considered to be isolated without a connection to a macrogrid.

Figure 6 indicates the proposed control system schematic for the base case.

The base case microgrid contains some flexibility in the way that loads and generation interact when battery energy storage is contained in the system. At any given time, the summation of the load and generation must be equal to avoid power quality issues and service interruption. The battery can supply energy in times of excess load, and absorb energy in times of excess generation, but only for a limited time. Once the battery is unable to meet the system's request to balance the load and generation, either generation or load must be curtailed, or additional load must come online to maintain stability.

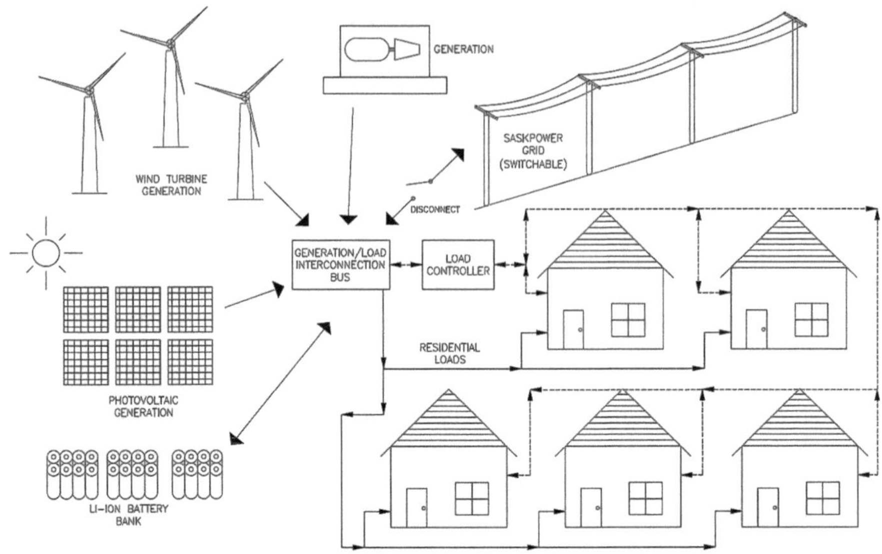

Fig. 5 Conceptual base case microgrid layout

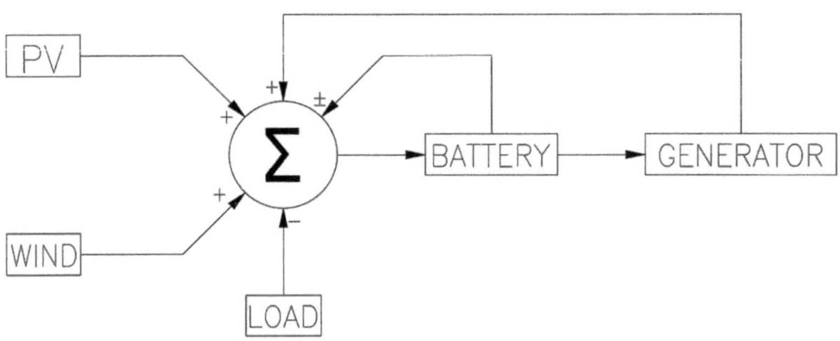

Fig. 6 Base case-control method diagram

It was determined early in the analysis that there was little benefit to the addition of wind turbine generation. With the cost of small wind power generation being \$6,280/kW, there was not enough financial advantage to including it in the renewable generation portfolio compared to the photovoltaic generation which was priced at \$4,170/kW. It was anticipated that photovoltaic generation being limited to daylight hours may be a detriment to the system at some point, whereas the ability to produce renewable energy at night with wind turbine generation may be a benefit and legitimize the presence of wind turbine generation in the optimized microgrid. It was, however, seen through the modelling that the fuel costs associated with night-time generation were low enough to negate the limited benefit of wind turbine generation. Wind generation was still assessed in various degrees of

Table 1 Lifetime costs for combinations of generation and storage without intelligence

Photovoltaic generation capacity rating (kW)							
		0	20	25	30	**35**	40
Energy storage capacity (kWh)	0	$581,517	$510,358	$517,201	$527,690	$540,537	$554,963
	10	$589,000	$502,672	$507,696	$516,921	$529,177	$543,315
	20	$596,484	$497,474	$499,409	$507,206	$518,301	$531,568
	30	$603,945	$494,652	$493,032	$498,575	$508,421	$520,971
	40	$611,427	$494,351	$488,685	$491,525	$499,635	$510,961
	50	$618,902	$496,716	$485,691	$485,784	$491,834	$501,912
	60	$626,377	$500,890	$484,794	$481,498	$485,488	$493,667
	70	$633,852	$506,521	$486,410	$478,648	$480,446	$487,194
	80	$641,327	$513,000	$489,842	$477,545	$477,024	$482,275
	90	$648,808	$520,125	$495,093	$479,086	**$475,366**	$479,314
	100	$656,271	$527,511	$501,349	$482,772	$475,872	$478,429

penetration in the microgrid for each analysis, however, in each scenario it was always more financially viable to exclude it from the generation portfolio. Furthermore, remote communities in Canada are predominantly located in the north, and a combination of cold temperatures and wind-driven snow and ice significantly limit the operation of wind turbines. Extremely cold temperatures are detrimental to the operation of mechanical systems, and icing of the turbine blades results in poor aerodynamic performance of the turbines.

Depending on the generation and storage combination, the propane generator must operate for a specific amount of time to ensure the load is met. The fuel cost associated with its operation is one of the major components of lifetime cost to the microgrid. Table 1 displays a portion of the calculated microgrid lifetime costs associated with varying amounts of battery energy storage, photovoltaic generation and associated propane generation costs. All costs are based on the previously mentioned costing information.

It can be seen that the optimal lifetime cost of the system occurred when the microgrid system consisted of 35 kW of photovoltaic generation and 90 kWh of battery energy storage. For greater amounts of photovoltaic generation there were diminishing returns as the increase in solar production could not be adequately contained for sufficient periods of time in the battery energy storage system. Likewise, if the size of the battery storage system was increased, the added storage would not be utilized often enough to justify the additional expenditure.

Note that when the microgrid has 0 kWh of storage, and 0 kW of photovoltaic generation, the lifetime cost is $581,517, or greater than 22 % higher than the optimized case with energy storage and a PV array.

This optimized combination of generation and storage is now used as a base case to determine the degree of benefit in adding a smart control system to the microgrid optimization model.

4 Application of a Smart Microgrid Management System to the Optimization Model

Algorithms are sets of instructions that contain logic with the capacity for executing calculations, processing data or enabling automated reasoning. Automated reasoning is a subfield of artificial intelligence and is associated with neural networks, fuzzy logic, expert systems and smart control systems. After a preliminary investigation, it was determined that although there may be sufficient value in the further exploration of neural networks, fuzzy logic and expert systems, the most immediate value proposition would be found in the use of smart control systems.

A smart control system is a closed-loop algorithm, as indicated in Fig. 7, which operates autonomously to achieve a specified outcome. For the purpose of this study, a smart control system will be employed to manage the generation components of the specified microgrid system. It will also be used to manage load shedding and load deferral. The intent was to create a viable business case for the inclusion of a smart control system in a microgrid system, and to prove this viability in the developed optimization model.

To facilitate the addition of intelligence to the system, opportunities for smart management were defined. Assumptions were made regarding the characteristics and components available in each residence. As previously indicated, it was found that the mean annual energy consumption for the monitored homes was 13,807 kWh and was comparable to the measured annual consumption of 14,000 kWh for a passive-solar, energy-efficient home built and monitored near Saskatoon which has its own well water supply and waste management system. It was acceptable then to assume that the acquired load profiles were indicative of a similarly designed home with similar characteristics.

The prime candidates for demand side management that were identified in the model were as follows: 0.355 kW of exterior lighting, a 50 US gallon electric hot water heater, and a 50 US gallon ground source heat pump (GSHP) mass tank.

Additional control could be added to the smart microgrid management system to automate load shedding and load deferral of other electrical loads in the homes. These loads could include televisions, microwaves, electric stoves, washing and drying machines, refrigerators, freezers, heating/air conditioning units and lighting.

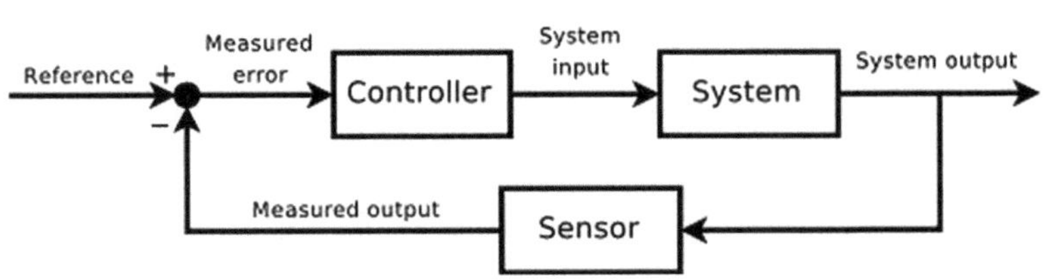

Fig. 7 Typical closed-loop system [22]

In the electrical code, all the aforementioned appliances except lighting and televisions must have dedicated circuits, thus using a smart control system to manage these systems would not be complicated, expensive or have detrimental effects on other components which could share the same circuit breaker. For this study, load curtailment was limiting to exterior lighting loads totaling 0.355 kW which was characteristic of the home being modeled and could be controlled during times when their use was not beneficial.

A 50 US gallon, two-element electric hot water heater is rated at 4.5 kW was also identified as a potential energy storage device. The temperature of a hot water heater is crucial to safety, and must be regulated between specific temperatures so as to decrease the probability of contracting legionella pneumophila pneumonia (also known as Legionnaires Disease) caused by domestic hot water temperatures below 49 °C, as well as to reduce the risk of tap water scalds which can be caused by temperatures above 60 °C [23]. The available heat energy stored in the hot water tank is given in Eq. (1).

$$\Delta Q = \frac{mc_p \Delta T}{3{,}600} \tag{1}$$

where

m Mass of water (189.27 kg)

c_p Specific heat water (4.18 J/g K)

ΔT Temperature differential (11 °C)

The energy storage capability of the hot water tank within the two temperature limits is 2.42 kWh per home, or a total of 12.1 kWh for the five-home system. This is the amount of energy storage available for energy management by the intelligent system. With an estimated hot water heater duty cycle of 12.5 % [24], the average power available from the hot water tank for curtailment when needed is given by Eq. (2). This is equivalent to using stored energy in the tank to meet the rest of the microgrid load.

$$P_{\text{avg HWH}} = (P_{\max})(D)(x) \tag{2}$$

where

$P_{\max}$ Maximum power draw of a single ground source heat pump, equal to 4.5 kW

D Ground source heat pump duty cycle, equal to 0.125

X Number of homes, equal to 5

Thus, the average hot water heater power available for deferral for the 5-home community is 2.8 kW based on the equation above. The hot water heater's operation in the smart microgrid will be modelled similar to a 2.8 kW battery with a total storage capacity of 12.1 kWh.

A 3 ton in-floor heat pump with a rating of 4.5 kW delivers heat to a 50 % ethylene glycol—water mixture which is stored in a 50 US gallon insulated mass tank in each home. This system is controlled by an integrated aquastat which monitors the temperature of the mass tank, and signals the in-floor heat pump to deliver heat when needed. The available temperature differential allowed in the mass tank is estimated to be 10 °C.

The available energy storage in the mass tank can be calculated as shown in Eq. (3).

$$\Delta Q = \frac{mc_p\Delta T}{3{,}600} \tag{3}$$

where

m Mass of 50 % ethylene glycol—water mixture, equal to 205.93 kg
c_p Specific heat of 50 % glycol—water mixture, equal to 3.41 J/g K
ΔT Temperature differential, equal to 10 °C

The resulting available energy for storage is 1.95 kWh/home. This equates to a cumulative energy differential of 9.75 kWh for a five-home network. The duty cycle of a GSHP heating system is 8.9 % based on measured energy consumption in the home, and the associated power rating was 4.5 kW/unit. This means that the average power from the geothermal mass tank energy storage in a five home system is given by Eq. (4).

$$P_{\text{avg GSHP}} = (P_{\max})(D)(x) \tag{4}$$

where

$P_{\max}$ Maximum power draw of a single ground source heat pump, equal to 4.5 kW
D Ground source heat pump duty cycle, equal to 8.9 %
X Number of homes, equal to 5

Thus, the average power available for deferral in the 5-home community contributed by the GSHP is 2.0 kW based on the equation above. Much like the case for the electric hot water heaters, this GSHP array will be modelled similar to a 2.0 kW battery with a total storage capacity of 9.75 kWh.

The smart microgrid management system was integrated with the optimization model in the MATLAB environment to create the smart microgrid optimization model and functions according to the following flow chart.

As described in Fig. 8, the optimization model will measure instantaneous load and identify loads available to be shed, then evaluate available energy generation from renewables and state of charge for the energy storage facilities. Next, it will determine if there is excessive generation thereby allowing the opportunity to charge energy storage devices, or if there is a negative generation balance thus requiring either discharge from energy storage devices, or necessitating the operation of the generator to operate the site and recharge electrical energy storage if there is sufficient capacity. Finally, the optimization model records data for the sample simulation run and calculates the cost of operation.

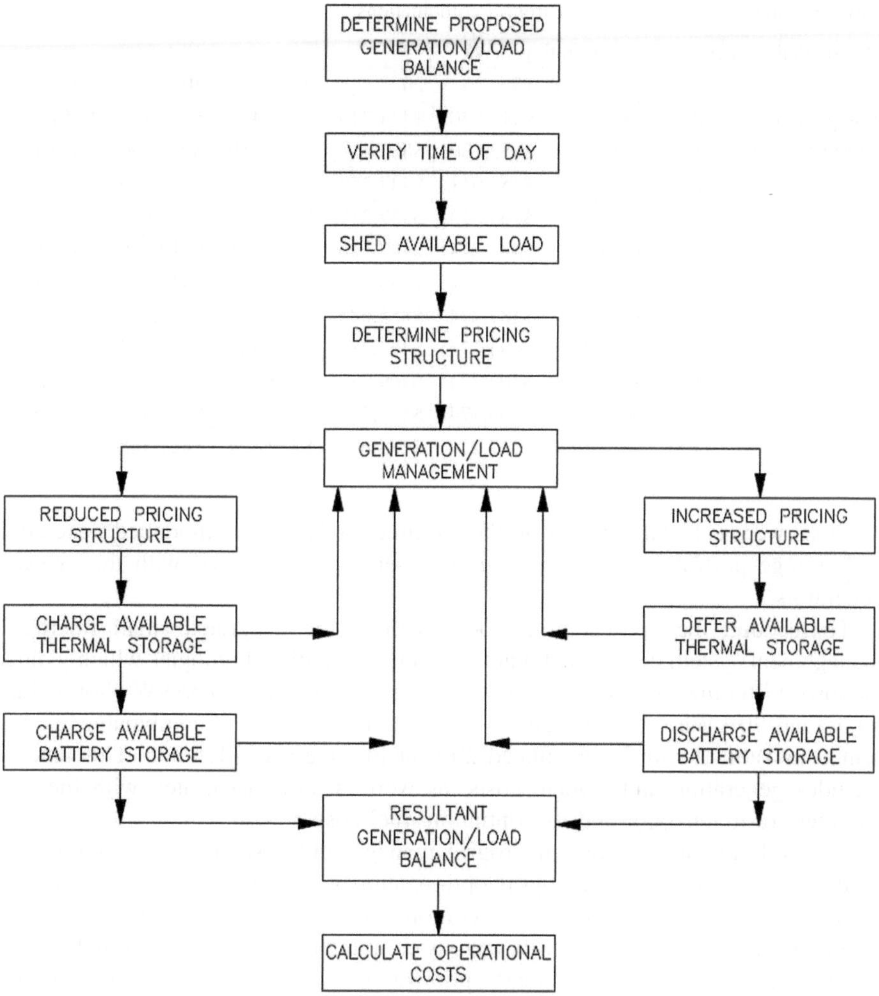

Fig. 8 Smart control algorithm flow chart

5 Results

The smart microgrid optimization model was used to analyze numerous combinations of generation and storage options ranging from 0 to 200 kW of each of wind turbine generation and PV generation in 5 kW increments, and 0 to 200 kWh of energy storage in 5 kWh increments, each time integrating thermal energy storage and load shedding techniques to determine the lifetime cost of the specific combination of systems.

Table 2 Lifetime costs of smart microgrid combinations

Photovoltaic generation capacity rating (kW)							
		0	15	20	25	**30**	35
Energy storage capacity (kWh)	0	$549,394	$447,930	$434,740	$434,443	$440,713	$451,494
	10	$556,871	$452,130	$433,462	$429,759	$433,830	$443,115
	20	$564,348	$457,911	$434,359	$425,934	$427,369	$434,913
	30	$571,829	$464,813	$437,389	$423,914	$422,654	$427,998
	40	$579,302	$472,274	$442,520	$424,241	$419,616	$423,004
	50	$586,775	$479,774	$448,739	$427,274	**$418,566**	$420,087
	60	$594,249	$487,274	$455,685	$432,223	$419,915	$419,317
	70	$601,725	$494,774	$462,785	$438,447	$423,684	$420,369
	80	$609,202	$502,274	$469,978	$445,156	$428,966	$423,286
	90	$616,687	$509,774	$477,274	$452,116	$434,986	$427,886
	100	$624,147	$517,274	$484,601	$459,126	$441,548	$433,582

Costs were calculated based on the required capital of the modelled generation and storage portfolio as well as ongoing fuel costs associated with the propane generators.

To recount, the optimization model without the integrated smart microgrid management system determined that the combination of technologies which yielded the lowest lifetime cost consisted of 0 kW of wind generation, a 35 kW photovoltaic array, and 90 kWh of battery energy storage. This optimized combination is presented in Table 2 and the calculated 20 year lifetime was $475,366. This amount includes generation and storage costs, as well as costs associated with the procurement of a generator and its continuing fuel costs.

The addition of the smart microgrid management system to the optimization model yielded the smart microgrid optimization model which resulted in both a reduction of required renewable generation capacity and battery energy storage. The lifetime costs associated with various combinations of generation and storage are indicated in Table 2. The smart optimization algorithm varied wind and solar generation from 0to 200 kW in 5 kW increments, and varied the battery storage from 0 to 200 kWh in 10 kWh increments.

Table 2 indicates the calculated 20-year lifetime costs of capital and operation of a microgrid operated by a smart microgrid optimization model which includes control of residential loads and thermal energy storage. The costs associated with the implementation of the smart microgrid management system have been included in the 20-year lifetime calculation considering a 5-home community with curtailable lighting and remote load control of both the electric hot water heater and the ground source heat pump mass tank.

The results from the smart microgrid optimization model indicated that the microgrid combination with the lowest lifetime cost consisted of a combination of 30 kW of PV generation and only 50 kWh of battery energy storage. The estimated lifetime cost of this combination was $418,566. In contrast, the optimized base case combination without smart microgrid management of internal thermal storage and load shedding yielded a total lifetime cost of $475,366. The smart microgrid optimization model yielded a combination of system components that led to a reduction of approximately 11.9 % in lifetime costs. As well, this means that the total savings of the smart microgrid optimization model compared to the initial base case of operating on fossil fuel-only yields savings of approximately 28.0 %

Approximately the same amount of PV generation was required for both the optimized systems with electrical energy storage and renewable generation, however, the smart microgrid option only required 50 kWh of battery energy storage, compared to the original base case which required 90 kWh. This is due, in large part, to the presence of internal thermal energy storage in the smart microgrid.

6 Power System Reliability Considerations

The studies carried out in the previous sections consider different combinations and capacities of power generation sources and energy storage facilities to meet the demand of the isolated microgrid in an acceptable manner. The level of reliability achieved with the different combinations can significantly vary, and a high level of reliability is usually associated with a relatively high system investment cost. Although the chief focus of this research project is to develop a smart microgrid controller that will reduce the cost of installing and operating a microgrid, it becomes important to determine the optimal system combination that provides reasonable and acceptable level of system reliability. Planning of small isolated power systems generally apply deterministic reliability methods that determine capacity reserve margins required above the peak load of the system. These methods, however, do not recognize random system behavior, and therefore, are not suitable to microgrids that are supplied by intermittent and uncertain generation from wind and photovoltaics. Probabilistic methods that can model stochastic system behavior are more suitable for such systems. This section presents the development of a probabilistic model to assess the reliability of the different system scenarios as discussed earlier, i.e. the base case system (generators only), the optimized base case system (generators and renewable energy) and the ideal microgrid (DSM, generators, renewable energy and energy storage). The developed model is verified against an existing reliability program to ensure accuracy.

There are two predominant methods of evaluating system reliability: analytical and simulation. To assess the proposed system's reliability, a simulation of the system was performed using Monte Carlo methods [25]. Monte Carlo methods simulate the actual process and functional operation of a system using random numbers as input.

The load and generation data employed to develop the most cost-effective combination of renewable generation, energy storage, load management and conventional generation were analyzed from a reliability perspective to substantiate the improved reliability of the lowest-cost combination of technologies as indicated by

Table 3 MTTF and MTTR parameters for the microgrid technologies

Technology	MTTF (h)	MTTR (h)	Availability (%)
Photovoltaic	119,902	36	99.97
Lithium ion energy storage	2,400	6	99.75
Propane generator	3,504	876	80.00

the smart microgrid optimization model. A Sequential Monte Carlo simulation (SMCS) was designed to estimate the reliability of the proposed system.

To verify the operation of the SMCS model, the microgrid base case without considering storage, demand side management or PV generation was first modeled in the developed computer program, and its reliability results were compared to the Small Isolated Power System Reliability (SIPSREL) program [26], a program developed by the University of Saskatchewan to determine the loss of load expectation (LOLE) for power systems. SIPSREL was not designed to determine the reliability of a system with both battery and thermal energy storage.

The reliability parameters, the mean time to failure (MTTF), and the mean time to repair (MTTR) were specified based on operator information for the PV and propane generation equipment, as well as the lithium ion battery energy storage system. Table 3 summarizes the MTTF and MTTR parameters used for the reliability calculation. These parameters are used as inputs to the SMCS program developed in this project.

It should be noted that both photovoltaic arrays and lithium ion energy storage systems largely consist of modular parts that can be easily replaced by non-trained persons. These components therefore have a relatively high reparability resulting in reduced downtime. The modular design also allows the systems to continue operating at derated states if part of the system fails. Although propane generators located close to service technicians have a much lower MTTR, the average time of 876 h stated in Table 3 is indicative of an isolated community which would require a technician to travel to site with parts to repair the generator.

Each iteration of the developed SMCS model simulated five years of load, and generation profiles and used MTTF and MTTR indices to determine generation 'up' and 'down' states throughout the simulation using Eqs. (5) and (6).

$$t_{\mathrm{up}} = -\mathrm{MTTF}\ln(\mathrm{Rand}) \tag{5}$$

$$t_{\mathrm{down}} = -\mathrm{MTTR}\ln(\mathrm{Rand}) \tag{6}$$

where

t_{up} Generator 'up' time
t_{down} Generator 'down' time
MTTF Mean time to failure
MTTR Mean time to repair
Rand Randomly generated number between 0 and 1

Convergence criteria were defined to determine an appropriate stop time for the SMCS runs. Each simulated iteration produced a unique LOLE. The deviation of the running average of the LOLE from the current iteration must be within a certain range of the running average in the previous iteration to establish convergence. The convergence criteria for this model were defined as a deviation not greater than within 0.1 % of the average LOLE for 50 consecutive iterations.

To verify the SMCS model, a base case system consisting of two propane generators was modelled using the SMCS model and the SIPSREL program. The LOLE of the microgrid with two propane generators met convergence criteria within 1358 iterations. After 1358 iterations, the LOLE of the system was 348.04 h/year. In comparison, the SIPSREL program yielded an LOLE of 351.32 h/year for the same system. This is a difference of only −0.93 %, which verified the accuracy of the developed SMCS model when propane generators were the only source of generation.

To further verify the operation of the SMCS model, the solar generation profile was incorporated into the system, and its results were compared to the SIPSREL program. Before the power system could be analyzed in SIPSREL, PV generation data had to be reduced to a generation table. To retain the seasonal and diurnal characteristics of the PV profile and its correlation to the load profile, PV generation data were separated into the following four categories:

1. Winter Day
2. Winter Night
3. Summer Day
4. Summer Night

Each day-time period consisted of the time from 8:00 am to 8:00 pm and the corresponding night-time period was from 8:00 pm to 8:00 am.

Sturges' Rule [27] was implemented to determine the least number of class intervals required to reduce the amount of load and generation data to be used in SIPSREL. Sturges' Rule is given in Eq. (7).

$$\mathrm{NoCl} = 1 + 3.3\,\log_{10}(\mathrm{N}) \tag{7}$$

where

NoCl Number of classes

N Number of data points considered, equal to 8735 in the winter and 8783 in the summer

Based on Sturges' Rule, the required number of classes to represent the PV generation profile for each of the proposed periods was 14.01, which was rounded up to 15 classes. Although Sturges' Rule indicated that each profile period could be reduced to 15 classes, the night-time PV generation data consisted mostly of '0' kW of generation. Hence, only five classes were used to represent night-time PV generation so as to maintain relevant generation data in each class. The class size for each class interval was evenly distributed over the generation range of the PV array.

Table 4 Summary of LOLE results

Configuration	LOLE (h/year)
Generators only	348.04
Generators and solar	249.83
Generators, PV and energy storage	171.39

According to SIPSREL, the LOLE of the microgrid consisting of two propane generators and a 35 kW solar array is 267.74 h/year. In comparison, the SMC model converged after 1,358 iterations and yielded an LOLE of 249.83 h/year, a difference of −6.69 %. Although this difference is higher than the generator-only model comparison, this is still within acceptable error considering the variability of introducing the solar model into the SIPSREL program with regards to the pre-processing that was required to reduce the PV generation to a 4-part generation table. Thus, the accuracy of the SMC model was once again verified.

Once the developed SMCS model was verified, sequential data for the operation of the optimized microgrid was entered into the model. Using the same convergence criteria, the model converged after 1,378 iterations and yielded an LOLE of 171.39 h/year. This is a decrease in LOLE of approximately 36 % compared to the same microgrid operating on only generators and solar power. This was as expected, and is the result of using a combination of load shedding and stored energy during times when the generation components were not online in order to maintain microgrid functionality. Table 4 shows that the LOLE values are significantly higher than acceptable values in large power systems. The improvement in reliability of isolated systems should, however, be justified by the associated costs to lower the LOLE values.

7 Conclusions

This chapter presented the methodology for developing a smart microgrid optimization model that can be used to determine the most financially economical combination of microgrid technologies with acceptable level of system reliability. The smart microgrid optimization model assessed various combinations of PV and wind turbine renewable generation in increments of 5 kW varying between 0 and 200 kW. Battery energy storage was also considered in combination with renewable generation in increments of 5 kWh from 0 kWh to 200 kWh. The smart microgrid optimization model was developed in the MATLAB environment and incorporated a smart microgrid management system to achieve additional fuel savings and increased system reliability by utilizing load shedding and load deferral techniques.

Microgrid power system reliability was assessed using an SMCS simulation model. The accuracy of the model was verified using the SIPSREL program developed at the University of Saskatchewan. Results indicated that the implementation of renewable generation, battery energy storage and DSM techniques can substantially reduce the lifetime operational costs of a microgrid while increasing power system reliability.

Implementation of energy storage technologies to reduce fossil fuel consumption as well as to increase renewable generation penetration is an area of ongoing research, as is the application of microgrid technology to improve remote power system reliability.

References

1. Benjamin K, Basso T, DeBlasio R (2008) Microgrid standards and technologies. In: Power and energy society general meeting-conversion and delivery of electrical energy in the 21st Century, 2008 IEEE. IEEE
2. Marnay C, Zhou, N, Qu M, Romankiewicz J (2012) Lessons learned from microgrid demonstrations worldwide. Ernest Orlando Lawrence Berkeley National Laboratory
3. Paulo Moisés C, Matos MA (2005) Reliability of distribution networks with microgrids. In: Power Tech, 2005 IEEE Russia. IEEE, 2005
4. Petrie EM, Lee Willis H, Takahashi M (2001) Distributed generation in developing countries. Cogeneration Onsite Power Prod 2.5(2001):41–49
5. O'Grady P (2012) Solar panel prices continue their downfall. Phoenix Bus J 29 Nov 2012
6. Off-Grid Communities. Aboriginal Affairs and Northern Development Canada. http://www.aadnc-aandc.gc.ca, Accessed Aug 2012
7. EIA, US (2013) Annual energy outlook 2013. US Energy Information Administration, Washington, DC
8. NRCan CA (2013) Introduction to smart micrognids in remote communities. Renewables in Remote Communities, Natural Resources Canada, Ottawa
9. Shafiqur R, Al-Hadhrami LM (2010) Study of a solar PV–diesel–battery hybrid power system for a remotely located population near Rafha, Saudi Arabia. Energy 35(12):4986–4995
10. Omar H, Bhattacharya K (2012) Optimal planning and design of a renewable energy based supply system for microgrids. Renew Energy 45:7–15
11. Weis TM, Ilinca A (2008) The utility of energy storage to improve the economics of wind–diesel power plants in Canada. Renew Energy 33(7):1544–1557
12. Roy B, Karki R (2001) Capacity expansion of small isolated power systems using PV and wind energy. IEEE Trans Power Syst 16(4):892–897
13. A-A Ali, Islam SM (2004) Demand side management for remote area power supply systems incorporating solar irradiance model. Renew Energy 29(13):2027–2036
14. Check-It Solutions Platform (2013) Solutions. http://www.check-it.ca/site/check-it-solutions-platform.html. Accessed April 2013
15. 16 kW Generac Generator. Digital image. Generac, n.d. Generac.com. Web. 25 April 2015
16. Specification Sheet. Generac Guardian Series 17/20 kW Residential Standby Generator. http://www.generac.com/Residential/GuardianSeries/17kW/, Accessed Jan 2012
17. Jansen R (2010) Photo of 8.19 kW Photovoltaic Array—Saskatoon." 2010. JPEG file
18. Raum 3.5 kW Wind Turbine. Digital image. Angel Wind Energy, n.d. angelwindenergy.com. Web. 25 April 2015
19. Vartanian C (2010) Energy storage and PV generation integration—utility and manufacturers perspectives. Proceedings of High Penetration Photovoltaics Workshop, Denver, CO
20. Boslet M (2012) Lithium battery prices on slow decline. http://www.greentechmedia.com/articles/read/lithium-battery-prices-on-slow-decline Accessed Jan 2012
21. MATLAB Version 7.5.0.342 (R2007b), The MathWorks Inc., Natick, MA, 2000

22. Dorf R, Modern control system. Addison-Wesley Series in Electrical Engineering, Boston, USA
23. Stanwick R (1986) Balancing the risks: legionella pneumophilia pneumonia and tapwater scalds in the home. Can Med Assoc J 135:1251–1252
24. How much does it cost to run water heater. Waterheatertimer. http://waterheatertimer.org/How-much-does-it-cost-to-ruun-water-heater.html. Accessed Dec 2011
25. Halton JH (1962) Sequential monte carlo. Math Proc Cambridge Philos Soc 58:57–78. doi:10.1017/S0305004100036227
26. Karki R, Billinton R (1997) SIPSREL (small isolated power system reliability software package) user's manual. University of Saskatchewan
27. Sturges HA (1926) The choice of a class interval. J Am Stat Assoc 21(153):65–66

Chapter 7
Renewable Energy Generation System Expansion Planning

Jaeseok Choi

Abstract This chapter titled "Renewable Energy Generation System Expansion Planning" proposes a new approach using linear programming to solve the long-term generation mix considering renewable energy resources generators namely; Wind Turbine Generators (WTGs) and Solar Cell Generator (SCGs) with multi-criteria considering CO_2, SO_x, and NO_x emission constraints under the uncertain circumstances. The chapter focuses on the development of tools to analyze the effect of WTGs and SCGs on the best generation mix (BGM) using fuzzy set theory. The method accommodates sensitivity analysis of capacity factors of WTG and SCG. The effectiveness of the proposed approach is demonstrated by applying to solve the flexible generation mix problem on the Korean power system for a planning horizon up to the year 2030. The test system contains nuclear, coal, LNG, oil, pumped-storage hydro plant, WTG and SCG in multi-years period. The economic credits of renewable energy resource in best generation mix problem were proposed in view point of CF and CER of WTG and SCG. It is noted that the WTG and SCG are competitive generator types in the mix and the WTG and SCG can win the conventional generators mix from for various percentage mix of WTGs and SCGs in the Korean power system. The proposed model and method are useful tools for various economics evaluation in BGM considering renewable energy generators and emission constraints.

1 Introduction

The use of renewable energy sources as a generation resource in electric power systems is growing rapidly due to its advantages that are minimal pollution, non-depletion and low operating cost.

J. Choi (✉)
Department of Electrical Engineering, Gyeongsang National University, Jinju, South Korea
e-mail: jschoi@nongae.gsnu.ac.kr

N.R. Karki et al. (eds.), *Sustainable Power Systems*, Reliable and Sustainable Electric Power and Energy Systems Management,
DOI 10.1007/978-981-10-2230-2_7

The use of wind turbine generator (WTG) and solar cell generator (SCG) are the fastest growing energy sources on a percentage basis. In the past 20 years, for example, the global wind energy capacity has increased five hundredfold—from 3.5 GW in 1994 to almost 50 GW by the end of 2004 and 1,560 GW by 2013. A major reason for this development is the incentive policies promoted by several countries to reduce the emission of pollutant gases stated by the Kyoto Protocol signed in 1997 in Japan and the Paris Agreement (COP21 UN) in France at Dec. 12, 2015. As a result, it is becoming more important to analyze the best generation mix including WTG and SCG. This is coupled with growing environmental awareness and increasing prospects ratification of the Kyoto Protocol and Paris Agreement (COP21 UN).

This chapter proposes a new approach to solve the long-term generation mix considering renewable energy resources generators namely; WTGs and SCGs with multi- criteria considering CO_2, SOx, and NOx emission constraints, under the uncertain circumstances is proposed using linear programming. This chapter focuses on the development tools to analyze the effect of WTGs and SCGs on the best generation mix (BGM) using fuzzy set theory. The method accommodates sensitivity analysis of capacity factors of WTG and SCG [27, 28]. The effectiveness of the proposed approach is demonstrated by applying to solve the flexible generation mix problem on the Korean power system for the year 2030. The test system contains nuclear, coal, LNG, oil, pumped-storage hydro plant, WTG and SCG in multi-years period.

2 The Concept of Flexible Planning

In recently, the uncertainties of external conditions for power system scheduling problem, demand growth, primary energy resource circumstances and reliability of energy supply system and energy security are becoming more and more unmanageable. The uncertainties require more flexible system for preventing power system from the unmanageable collapse. Figure 1 means that the flexible planning and operation for power system is one which, "Although not necessarily gives the optimum solution for the basic forecasted conditions, yet can keep the reasonable scheduling solution from being significantly worsened by any assumed changes in the surrounding situations" [19, 20].

The useful methods for flexible generators mix problem and generators maintenance flexible scheduling problem using fuzzy LP, DP, IP and search method have been developed by the authors. Some cases of application of fuzzy set theory for the flexible planning and operation of power systems are introduced in this seminar. It is expected that more flexible solution can be obtained with the proposed methods because the fuzzy set theory that can reflect the subjective decision of decision-maker is used in these studies. This chapter uses fuzzy set theory in order

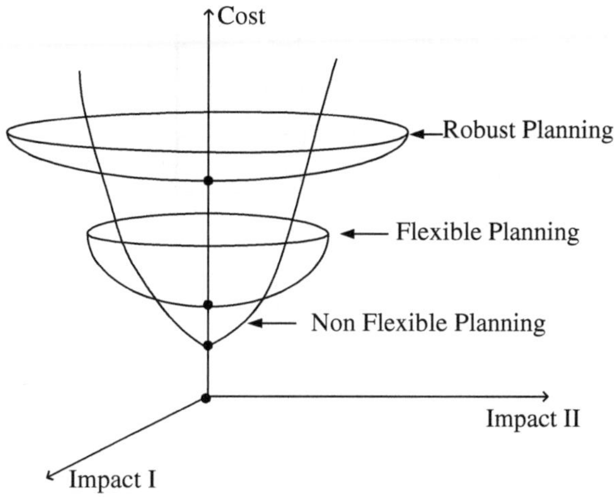

Fig. 1 Concept of flexibility

to get a flexible long-term generation mix solution in Korea power system considering CO_2 emission constraint and renewable energy generators as like as WTG (Wind Turbine Generator) and SCG (Solar Cell Generator).

3 The LP Formulation of Best Generation Mix

3.1 Problem Statement

This BGM problem is to determine the generation mix under the following assumptions:

1. The annual loads are known.
2. The number of generator is not that of units but that of types.
3. Nuclear power plants are able to perform load following.
4. It is assumed that the yearly capacity factors (CFs) of WTG and SCG are given.

For the proposed method, the system can be modeled as shown in Fig. 2. It is considered to be composed of Nuclear = 1, Coal = 2, LNG = 3, Oil = 4, P-G (pumped generator hydro) = 5, WTG = 6, SCG = 7 in convenient in this chapter.

Because wind speed and solar radiation do not maintain a specified stable level and the resource supply cannot be controlled by operator, the WTG and SCG have resource supply uncertainties rather than generator outage. Figure 3 shows the uncertainties of WTG and SCG plants are different from those of conventional plants. Therefore, the CFs of the WTG and SCG depend on the resource supply entirely with an associated ambiguity. In this chapter, it is assumed that the CFs are available. Therefore, the energy based on the specified CF should be consumed mandatory as is the case with energy limited generator and self-flow hydro generator.

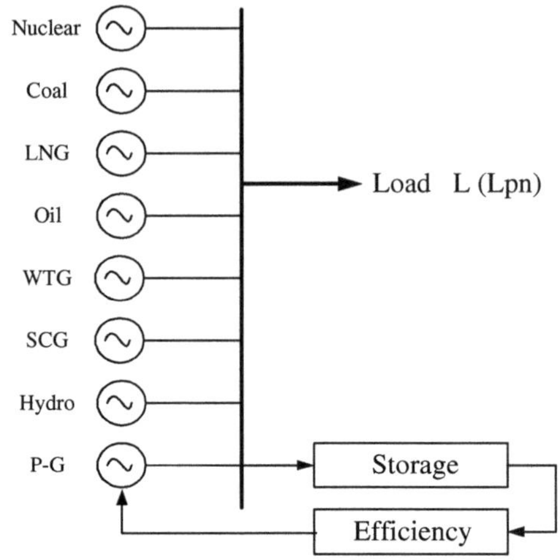

Fig. 2 A system model for use by the proposed method

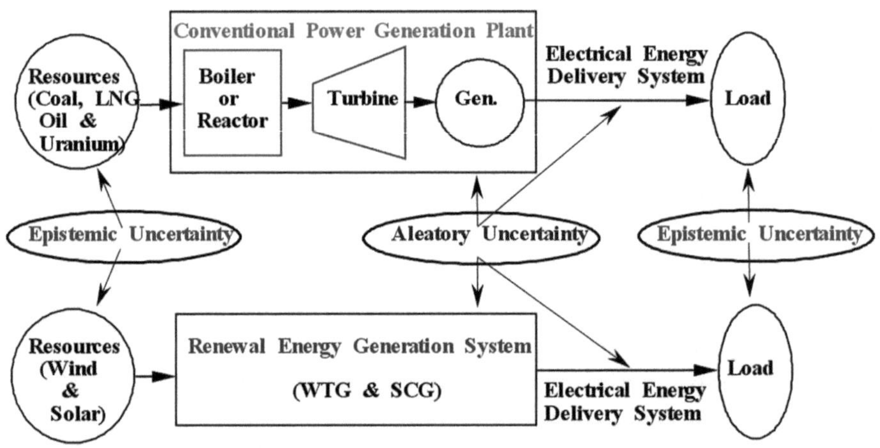

Fig. 3 Uncertainties of the renewable energy power and conventional power plants

In this study, it is assumed that the hydro generator construction is separately planned from that of the other kinds of generating units. In actual systems, the basic resources, which are reserves, of the hydropower plants have limitation in the country. Therefore, the choice of hydro plant construction is not much and non-flexible. Under these assumptions, the best generation mix problem is formulated as follows.

3.2 Objective Functions

3.2.1 The Economic Criterion

The economic criterion in the BGM is to minimize the sum of the construction cost and the fuel cost as

$$\begin{aligned}\text{Minimize } Z &= \sum_{n=1}^{N}\sum_{i=1}^{NG} K_{\text{cin}} d_{\text{in}} \alpha_i \Delta x_{\text{in}} + \sum_{n=1}^{N}\sum_{i=1}^{NG} K_{\text{fin}} f_{\text{in}} y_{\text{in}} \\ &= F(\Delta x_{\text{in}}, y_{\text{in}})\end{aligned} \quad (1)$$

where,

i	Unit type number (1 for nuclear, 2 for coal, 3 for LNG, 4 for oil and 5 for pumped-storage generators are specified in this chapter)
N	Number of total study stage year
NG	Number of unit type
	$K_{\text{cin}} = ((1 + e_{ci})/(1 + r))^{n\Delta Tk,k-1}$
	$K_{\text{fin}} = ((1 + e_{fi})/(1 + r))^{n\Delta Tk,k-1}$
e_{ci}	Apparent escalation rate of construction materials of unit
e_{fi}	Apparent escalation rate of fuel of unit
r	Discount rate
$\Delta T_{k,k-1}$	Step size years of study years from k to k−1
d_{in}	Construction cost of unit in year n
f_{in}	Marginal fuel cost of unit in year n (Won/MWh)
α_i	Annual expenses rate of unit
Δx_{in}	Construction capacity of unit in year n (MW)
y_{in}	Generation capacity of unit in year n (MWh)

3.2.2 Constraints

1. Installed capacity constraint

   where

R_n Supply reserve rate in n year (p.u)

$$\sum_{i=1}^{NG} (x_{\text{in}-1} + \Delta x_{\text{in}}) + \text{HYD} \geq L_{\text{pn}}(1 + R_n) \quad n = 1 \sim N \quad (2)$$

HYD_n Capacity of hydro generator in year n. In this study, it is assumed that the HYD_n is given

2. Energy constraint of demand

$$\sum_{i=1}^{NG} y_{in} \geq (L_{pn} + L_{Bn}) \times 8760/2 + V_n - \mathrm{HYD}_n \times 8760 \times \mathrm{CF}_H \quad (3)$$

where

L_{Pn} Peak load at n year

L_{Bn} Base load at year n

V_n The added demand energy is caused by pumped-storage generator

$\mathrm{CF_H}$ Average capacity factor of hydro generator

3. Production energy constraint of generation system

$$\begin{aligned} y_{in} &\leq (x_{in-1} + \Delta x_{in}) \times 8760 \times \mathrm{CF}_i \quad i = 1 \sim 5, \quad n = 1 \sim N \\ y_{in} &= (x_{in-1} + \Delta x_{in}) \times 8760 \times \mathrm{CF}_i \quad i = 6 \text{ and } 7, \quad n = 1 \sim N \end{aligned} \quad (4)$$

where

CF_i Average capacity factor of the i-unit

4. Capacity constraint in the initial year

$$x_{io} = \mathrm{EX}_i \quad i = 1 \sim \mathrm{NG} \quad (5)$$

where

EX_i Capacity of the existing unit

5. Constraint of mutual relationship between existing generator capacity and new generator capacity (state equation)

$$x_{in} = x_{in-1} + \Delta x_{in} \quad i = 1 \sim \mathrm{NG},\ n = 1 \sim N \quad (6)$$

6. Energy constraint of LNG thermal plant

$$y_{3n} \geq \mathrm{LEP}_{\min_n} / \rho_3 \quad n = 1 \sim N \quad (7)$$

where

$\mathrm{LEP}_{\min_n}$ LNG thermal generator production energy for LNG minimum due to consumption in n year

7. Constraints of reservoir capacity of pumped-storage generator

$$y_{5n} = \eta_g \times V_n \tag{8}$$

where

η_g Efficiency of pumped-storage generator

8. No load following power constraints of nuclear power plant

$$(x_{1n} - x_{5n}) \leq L_{\mathrm{Bn}} \quad n = 1 \sim N \tag{9}$$

9. Upper-Lower constraints of new unit capacity

$$\Delta X_{\mathrm{min_{in}}} \leq \Delta x_{\mathrm{in}} \leq \Delta X_{\mathrm{max_{in}}} \quad i = 1 \sim \mathrm{NG},\ n = 1 \sim N \tag{10}$$

where, $X_{\mathrm{max_{in}}}$ and $X_{\mathrm{max_{in}}}$ are the minimum and the maximum capacity of new units at year n respectively.

10. CO_2 constraint

$$\sum_{i=1}^{\mathrm{NG}} \xi_i y_i \leq \mathrm{CO}_{2\mathrm{MAX}_n} \tag{11}$$

where

ξ_i Fuel consumption rate of the *i*th unit (Ton/MWh)

$\mathrm{CO}_{2\mathrm{MAX}_n}$ Maximum quantity of CO_2 permitted in year n (Ton/year)

11. $\mathrm{SO_X}$ constraint

$$\sum_{i=1}^{\mathrm{NG}} \mathrm{SO}_{X_{\mathrm{in}}} \xi_i y_n \leq \mathrm{SO}_{\mathrm{XMAX}_n} \tag{12}$$

where

$\mathrm{SO}_{X_{\mathrm{in}}}$ $\mathrm{SO_X}$ density of the *i*-unit in year n (ppm/Ton)

$\mathrm{SO}_{\mathrm{XMAX}_n}$ Maximum quantity of $\mathrm{SO_X}$ permitted in year n (ppm/year)

12. $\mathrm{NO_X}$ constraint

$$\sum_{i=1}^{\mathrm{NG}} \mathrm{NO}_{X_{\mathrm{in}}} \xi_i y_n \leq \mathrm{NO}_{\mathrm{XMAX}_n} \tag{13}$$

where

$NO_{X_{in}}$ NO_X density of the *i*-unit in year n (ppm/Ton)

NO_{XMAX_n} Maximum quantity of NO_X permitted in year n (ppm/year)

4 The Fuzzy LP Formulation of Flexible Generation Mix

4.1 The Optimal Decision Theory by Fuzzy Set Theory

The fuzzy decision D resulting from q fuzzy goals $G_1, \ldots, G_q$ and p fuzzy constraints $C_1, \ldots, C_p$ is the intersection of them;

$$D = \left(\bigcap_{i=1}^{q} G_i\right) \cap \left(\bigcap_{j=1}^{p} C_i\right) \tag{14}$$

And also its membership function μ_D resulting from fuzzy goals and constraints is defined by

$$\mu_D(x) = \min\left[\min_{i=1\sim p} \mu_{G_i}, \min_{j=1\sim q} \mu_{C_i}\right] \tag{15}$$

where, min is an abbreviation of minimum. The fuzzy mathematical programming problem consists of finding the maximum of the fuzzy decision D.

$$\mu_D(x^*) = \max \mu_D(x) \tag{16}$$

Where $\mathbf{x}^*$ is the optimal decision solution max is an abbreviation of maximum. The vector Eq. (16) can be rewritten as the Eq. (17).

$$\mu_D(x_1^*, x_2^*, \ldots, x_N^*) = \max_{x_1 \cdots x_N} \mu_D(x_1, x_2, \ldots, x_N) \tag{17}$$

Figure 4 shows the concept for optimal decision on fuzzy sets, which is proposed by Bellman Zadeh.

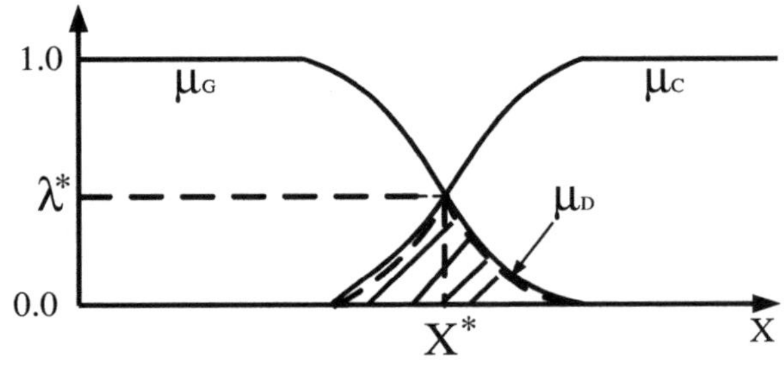

Fig. 4 Concept for optimal decision on fuzzy sets

4.2 The Function of Fuzzy Linear Programming

Fuzziness of Cost: $Z \lessapprox Z_0$

The Satisfaction Level in the best generation mix is flexible because the total cost is unjustly.

Fuzziness of Reliability criterion: $R_n \gtrapprox R_{0n}$

The reliability criterion in the long term based best generation mix may be flexible criterion than entirely strict. Using parameter λ, therefore, which means satisfaction level physically, best generation mix problem can be formulated as following.

Objective functions: Maximize λ

Constraints Subject to $F\ (\Delta x_{\text{in}}, y_{\text{in}}) + \Delta Z_0 \lambda = Z_0 + \Delta Z_0$

$$\left(\left(\sum_{i=1}^{NG} (x_{\text{in}-1} + \Delta x_{\text{in}}) + \text{HYD}_n\right) - L_{\text{pn}}\right) / L_{\text{pn}} - \Delta R_o \lambda = R_o - \Delta R_o$$

$$\sum_{i=1}^{NG} y_{\text{in}} \geq (L_{\text{pn}} + L_{\text{Bn}}) \times 8760/2 + V_n - \text{HYD} \times 8760 \times \text{CF}_H$$

$y_{\text{in}} \leq (x_{\text{in-1}} + \Delta x_{\text{in}}) \times 8760 \times \text{CF}_i$

$x_{i0} = \text{EX}_i$

$x_{\text{in}} = x_{\text{in-1}} + \Delta x_{\text{in}}$

$y_{3n} \geq \text{LEP}_{\text{min}n}/\rho_3$

$y_{5n} = \eta_g \text{x} Vn$

$(x_{1n} - x_{5n}) \leq \text{L}_{\text{B}n}$

$\Delta x_{\text{in}} \leq \text{DX}_{\text{max}_{\text{in}}}$

$$\sum_{i=1}^{NG} \xi_i y_i \leq \text{CO}_{2\text{MAX}_n}$$

$$\sum_{i=1}^{NG} \text{SO}_{X_{\text{in}}} \xi_i y_n \leq \text{SO}_{\text{XMAX}_n}$$

$$\sum_{i=1}^{NG} \text{NO}_{X_{\text{in}}} \xi_i y_n \leq \text{NO}_{\text{XMAX}_n}$$

5 The Fuzzy LP Formulation of Flexible Generation Mix

Budget for long-term generation expansion planning is given unjustly and usually in exact. As it is, it is fuzzy. The optimal decision theory by fuzzy set theory is a reasonable methodology for solving the optimal problem including the multiple objectives and ambiguity. The theory is introduced in detail in Appendix. Aspiration level for decision makers may be expressed by (18). In addition, the reliability criterion in the long term based best generation mix may be a flexible criterion rather than entirely strict as in (19).

$$\text{Fuzziness of Cost} : Z \lesssim Z_0 \tag{18}$$

$$\text{Fuzziness of Reliability criterion} : R_n \gtrsim R_{0n} \tag{19}$$

where

Z_0 Aspiration level of budget

R_{0n} Aspiration level of reliability criterion in n-year

Using parameter λ, which refers to the term satisfaction level, the best generation mix problem can be formulated as follows.

Objective functions: Maximize λ

Constraints Subject to $F(\Delta x_{in}, y_{in}) + \Delta Z_0 \lambda = Z_0 + \Delta Z_0$

$$\left(\left(\sum_{i=1}^{NG} (x_{in-1} + \Delta x_{in}) + \text{HYD}_n\right) - L_{pn}\right) / L_{pn} - \Delta R_o \lambda = R_o - \Delta R_o$$

$$\sum_{i=1}^{NG} y_{in} \geq (L_{pn} + L_{Bn}) \times 8760/2 + V_n - \text{HYD} \times 8760 \times \text{CF}_H$$

$y_{in} \leq (x_{in\text{-}1} + \Delta x_{in}) \times 8760\text{xCF}_i$

$x_{i0} = \text{EX}_i$

$x_{in} = x_{in\text{-}1} + \Delta x_{in}$

$y_{3n} \geq \text{LEP}_{minn}/\rho_3$

$y_{5n} = \eta_g \times Vn$

$(x_{1n} - x_{5n}) \leq \text{L}_{Bn}$

$\Delta x_{in} \leq \text{DX}_{max_{in}}$

$$\sum_{i=1}^{NG} \xi_i y_i \leq \text{CO}_{2\text{MAX}_n}$$

$$\sum_{i=1}^{NG} \text{SO}_{X_{in}} \xi_i y_n \leq \text{SO}_{\text{XMAX}_n}$$

$$\sum_{i=1}^{NG} \text{NO}_{X_{in}} \xi_i y_n \leq \text{NO}_{\text{XMAX}_n}$$

6 Case Studies

The generation mix was performed on Korean power system to demonstrate the effectiveness of the proposed method. The test system consists of 8 types generators over 22 years (2008–2030) with the initial year being 2008. The maximum, minimum load and the hydro capacity in standard years are listed in Table 1. The characteristics and economic data are summarized in Table 2.

Maximum new construction capacity is limited by limitation of human and equipment resources, etc. It is assumed, as shown in Table 3, that the maximum and minimum capacities per a stage year of new generator are same for all stage/year in this study. Table 4 shows maximum permissible limitation of air pollution emission input data.

Table 1 Maximum load, minimum load, and hydro plant at standard years

Years	Peak load L_P (MW)	Base load L_B (MW)	Hydro (MW)	LNG (10^3 Ton)
2008	58,990	35,394	1,800	–
2013	65,940	39,564	2,000	4,500
2018	70,050	42,030	2,200	5,500
2023	74,000	44,400	2,400	6,500
2028	77,000	46,200	2,600	7,500
2030	80,000	48,000	2,600	7,500

Table 2 The characteristics and economic data

Gen. type	Initial capacity (MW)	Fixed charge (10^3 Won/kW)	[a]A_{ER} of fixed charge (%)	Marginal fuel cost (Won/kWh)	[a]A_{ER} of fuel cost (%)
Nucl.	17,716	2,122	3.5	3.730	1
Coal	18,465	1,145	3.4	17.794	1
LNG	17,437	741	3.3	65.327	1
Oil	4,686	1,881	3.3	86.987	4
P-G	3,300	915	3.5	0	0
WTG	20	4,770	−4.5	0	0
SCG	10	7,500	−6.5	0	0

Annual cost rate (%)	Capacity factor (%)	Fuel consumption rate (Ton/MWh)	Density (ppm/Ton) CO_2, SO_X, NO_X
19	90	–	
17	90	0.2300	−450–500
17	60	0.1100	−200–300
17	55	0.2000	−200–100
13	30	–	
10	10	0	
10	15	0	

[a]A_{ER} means the apparent escalation rate (CER) and the discount rate is assumed as 7.5 %

Table 3 Maximum and minimum of capacity per a stage year of new generators (MW)

Generation type	ΔX_{max}	ΔX_{min}
Nuclear	5,000	0
Coal	5,000	0
LNG	5,000	30,000 × 5.25[a]
Oil	500	0
P-G	1,000	0
WTG	500–3,000	100
SCG	500–3,000	100

[a]The limited energy of LNG

Table 4 Maximum permissible limitation of air pollution emission (CO_2:[10^6 Ton/year], SO_X and NO_X:10^9[ppm/year])

Air pollution	2013	2018	2023	2028	2030
CO_{2MAX}	40	40	40	40	40
SO_{XMAX}	40	40	40	40	40
NO_{XMAX}	40	40	40	40	40

Table 5 Best generation mix considering renewable energy generators using the proposed method (%)

Gen. type	Base case: Case F1					
	2008	2013	2018	2023	2028	2030
Nucl	27.93	29.40	31.83	33.64	36.65	36.68
Coal	29.11	30.37	32.02	30.61	28.93	28.82
LNG	27.49	25.19	22.35	20.01	18.91	18.84
Oil	7.39	6.07	5.38	4.82	4.55	4.54
P-G	5.20	5.57	4.94	4.42	4.18	4.16
WTG	0.03	0.67	0.71	0.74	0.80	0.89
SCG	0.02	0.14	0.24	3.30	3.47	3.55
Hyd	2.84	2.59	2.53	2.47	2.53	2.52
Total	100	100	100	100	100	100

6.1 Test Results

The proposed fuzzy set theory based best generation mixes were simulated considering WTG and SCG and air pollution criterion. Aspiration level (Z_0) of total cost (budget) was specified as 27,500 Billion Won. The BGM result is shown in Table 5. The satisfaction level for the BGM was found to be 0.832. The result indicates an increase in the mix of nuclear power plants and renewable energy generators increase and a decrease in that of the coal power plants. Figure 5 is a graphical representation of the percent ratio results.

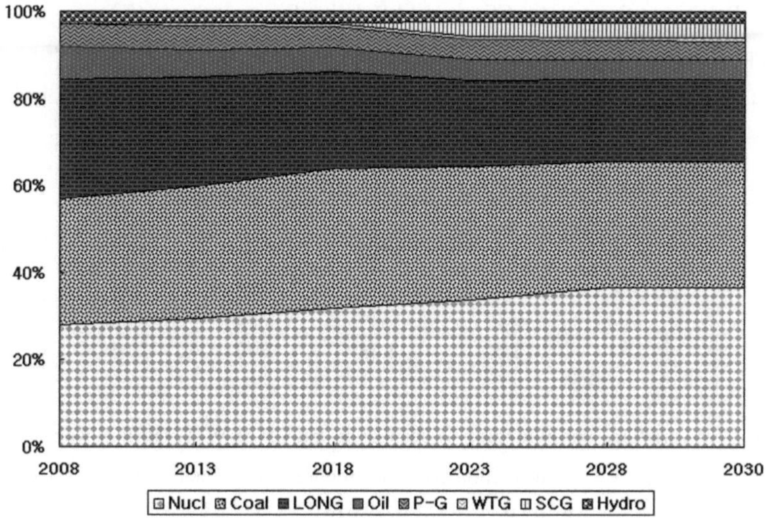

Fig. 5 Best generation mix by the proposed method (Case F1)

6.2 Sensitivity Analysis

6.2.1 Capacity Factor of WTG and SCG

The CF of WTG and SCG in BGM is important in order to determine the mix of WTG and SCG because they can be evaluated more reasonable by higher accurate CF of the WTG and SCG. The CF of WTG and SCG depend on the resources which are wind speed and solar radiation respectively. The capacity factor includes ambiguity. The capacity factors of WTG and SCG are usually in the ranges 8–16 % and 14–18 % respectively. Tables 6 and 7 show the best generation mix results at CF_{WTG} = 12 % and CF_{WTG} = 14 % respectively. It is noted that the WTG is competitive with SCG at CF_{WTG} = 14 % as shown in Table 7. Tables 8 and 9 show the cost evaluation results for the various CF of WTG and SCG.

Table 6 Best generation mix for CF_{WTG} = 12 % using fuzzy set theory (%)

Gen. type	Case CF3 (CF_{WTG} = 12 %)					
	2008	2013	2018	2023	2028	2030
Nucl	27.93	29.39	31.82	33.61	36.71	36.76
Coal	29.11	30.36	32.02	30.60	28.99	28.88
LNG	27.49	25.21	22.37	20.02	18.96	18.89
Oil	7.39	6.06	5.38	4.81	4.56	4.54
P-G	5.20	5.56	4.94	4.42	4.19	4.17
WTG	0.03	0.67	0.71	0.78	0.84	0.93
SCG	0.02	0.14	0.24	3.30	3.22	3.31
Hyd	2.84	2.59	2.53	2.47	2.53	2.52

Table 7 Best generation mix for CF_{WTG} = 14 % Using fuzzy set theory (%)

Gen. type	Case CF4 (CF_{WTG} = 14 %)					
	2008	2013	2018	2023	2028	2030
Nucl	27.93	29.39	31.82	33.61	36.71	36.76
Coal	29.11	30.34	31.78	30.56	28.95	28.84
LNG	27.49	25.26	22.15	20.04	18.99	18.92
Oil	7.39	6.06	5.31	4.81	4.55	4.54
P-G	5.20	5.56	4.88	4.41	4.18	4.16
WTG	0.03	0.67	1.72	3.85	3.74	3.82
SCG	0.02	0.14	0.24	0.32	0.40	0.49
Hyd	2.84	2.59	2.53	2.47	2.53	2.52

Table 8 Cost variations corresponding to changes in the CF of WTG (Billion won)

	CF_{WTG} (%)	Construction cost	Operating cost	Total cost	Satisfaction level
Case CF2	8	8,355.0	19,618.5	27,973.5	0.828
Case CF1	10	8,417.9	19,543.8	27,961.8	0.832
Case CF3	12	8,420.9	19,519.5	27,940.4	0.840
Case CF4	14	8,458.4	19,437.5	27,895.9	0.856
Case CF5	16	8,565.0	19,259.0	27,824.0	0.882

Table 9 Cost variations corresponding to changes in the CF of SCG (Billion won)

	CF_{SCG} (%)	Construction cost	Operating cost	Total cost	Satisfaction level
Case CF6	14	8,444.9	19,541.4	27,986.3	0.823
Case CF1	15	8,417.9	19,543.8	27,961.8	0.832
Case CF7	16	8,396.5	19,541.7	27,938.2	0.840
Case CF8	17	8,352.5	19,563.9	27,916.5	0.849
Case CF9	18	8,693.8	19,186.1	27,879.9	0.862

From the results, as the CF of the renewable energy generators increases, the total cost decreases because the operating cost of conventional generator decreases relatively even if the construction cost increases. The total costs per unit CF decreased from the higher CFs of WTG and SCG yield about 20 Billion Won/% and 27 Billion Won/% respectively. The result describes the economic credit of renewable energy resource in BGM problem. Finally, the higher CF makes the satisfaction level to increase because the total cost is decreased. Equation (2) Apparent escalation rate (e_c) of construction materials of WTG and SCG

It is expected that the apparent escalation rate (CER, e_c) of construction material cost of the WTG and SCG will be decreased as a result of technology development that has been taking place. Therefore, the negative CER may be considered. Specially, the CER of SCG will be decreased more rapidly rather than the CER of the WTG. For example, Tables 10 and 11 show the best generation mix result at

Table 10 Best generation mix for $e_{cSCG} = -7.0$ % using the proposed approach (%)

Gen. type	Case F12 ($e_{cSCG} = -7.0$ %)					
	2008	2013	2018	2023	2028	2030
Nucl	27.93	29.38	31.83	33.61	36.71	36.77
Coal	29.11	30.35	31.97	30.61	29.00	28.89
LNG	27.49	25.76	22.87	20.46	19.38	19.31
Oil	7.39	6.06	5.38	4.81	4.56	4.54
P-G	5.20	5.56	4.94	4.42	4.19	4.17
WTG	0.03	0.16	0.25	0.33	0.41	0.50
SCG	0.02	0.14	0.24	3.30	3.22	3.31
Hyd	2.84	2.59	2.53	2.47	2.53	2.52

Table 11 Best generation mix for $e_{cSCG} = -5.5$ % using the proposed approach (%)

Gen. type	Case F10 ($e_{cSCG} = -5.5$ %)					
	2008	2013	2018	2023	2028	2030
Nucl	27.93	29.43	31.85	34.85	37.62	38.33
Coal	29.11	30.39	32.03	31.24	30.22	29.79
LNG	27.49	25.14	22.31	20.68	19.37	19.09
Oil	7.39	6.07	5.38	4.99	4.67	4.61
P-G	5.20	5.57	4.94	4.58	4.29	4.23
WTG	0.03	0.67	0.71	0.77	0.82	0.90
SCG	0.02	0.14	0.24	0.33	0.41	0.50
Hyd	2.84	2.59	2.53	2.47	2.53	2.52

Table 12 Total cost evaluation corresponding to changes in the CER of WTG (Billion Won)

	e_{cWTG} (%)	Construction cost	Operating cost	Total cost	Satisfaction level
Case F10	−3.5	8,360.6	19,610.7	27,971.3	0.829
Case F11	−4.0	8,423.6	19,543.8	27,967.5	0.830
Case F1	−4.5	8,417.9	19,543.8	27,961.8	0.832
Case F12	−5.0	8,412.5	19,543.8	27,956.3	0.834
Case F13	−5.5	8,407.3	19,543.8	27,951.1	0.836

$e_{cSCG} = -7.0$ % and $e_{cSCG} = -5.5$ %. Tables 12 and 13 show the cost results for the various CER of WTG and SCG. As the CER of renewable generators increases, the total cost decreases because the construction cost of renewable energy generator decreases regardless of whether the operating cost of conventional generators increases or stays fixed. It is noted that the WTG and SCG can win conventional generators in mix from $e_{cWTG} = -4.0$ % and $e_{cSCG} = -6.0$ % respectively as shown in Tables 10, 11, 12 and 13. The total costs per unit ER decreased by the lower CER of WTG and SCG yield about 10 Billion Won/% and 50 Billion Won/%.

Table 13 Total cost evaluation corresponding in changes in the CER of SCG (Billion Won)

	e_{cSCG} (%)	Construction cost	Operating cost	Total cost	Satisfaction level
Case F10	−5.5	8,385.6	19,613.4	27,999.0	0.818
Case F11	−6.0	8,454.4	19,534.1	27,988.5	0.822
Case F1	−6.5	8,417.9	19,543.8	27,961.8	0.832
Case F12	−7.0	8,391.8	19,543.8	27,935.6	0.842
Case F13	−7.5	8,372.4	19,534.8	27,907.2	0.852

The results describe the economic credits of renewable energy resources in view point of construction material cost. The lower escalation rate makes the satisfaction level increase because total cost is decreased.

7 Conclusions

This chapter has proposed a fuzzy LP-based approach for the long-term best generation mix with multi-criteria considering renewable energy generator such as WTGs and SCGs and air pollution constraints. The proposed method can accommodate sensitivity analysis of CF and CER of WTG and SCG including uncertainty.

The effectiveness of the proposed approach is demonstrated by applying it to the best multi-year generation mix problem of the Korean power system for year 2030. The test system contains nuclear, coal, LNG, oil, pumped-storage hydro, WTG and SCG plants. The economic credits of renewable energy resource in best generation mix problem were proposed in view point of CF and CER of WTG and SCG. It is noted that the WTG and SCG are competitive generator types in the mix and the WTG and SCG can win the conventional generators in mix from $e_{cWTG} = -4.0$ % and $e_{cSCG} = -6.0$ % respectively for the Korean power system. Work a technique for a more accurate evaluation of the CF is under going and results will be reported as soon as they become available. The proposed model and method are useful tools for various economics evaluation in BGM considering renewable energy generators and emission constraints.

References

1. Covarrubias AJ, Expansion Planning for electric power systems. IAEA Bull 21(2/3):55–64
2. Chuang AS, Wu F, Varaiya P, A game-theoretic model for generation expansion planning: problem formulation and numerical comparisons
3. Zhu J, Chow M (1997) A review of emerging techniques on optimal generation expansion planning. In: IEEE trans on power systems, vol 12, no 4, Nov 1997
4. Das PK, Chanda RS, Bhattacharjee PK, Combined generation and transmission system expansion planning using implicit enumeration and linear programming technique. IE(I) J –EL 110–116

5. Climaco J, Henggeler Antunes C, Gomes Martins A, Traca Almeida A (1995) A multiple objective linear programming model for power generation expansionplanning. Int J Energy Res 19:419–432
6. Chung TS, Li YS, Wang ZY (2004) Optimal generation expansion planning via improved genetic algorithm approach. Electr Power Energy Syst 26:655–659
7. Park J, Jeong S, Choi J, Yun Y, Cha J, Yoon Y (2008) A Study on the relationship between probabilistic and deterministic reliabilities in Korea. PMAPS2008, Rincón, Puerto Rico, 25–29 May 2008
8. Choi J, Lee S, Il Kang K, Song K, Namkung JY (1994) The construction of the multi-criteria generation mix by the Fuzzy dynamic programming. In: Proceedings of international conference on neural information proceeding, '94-Seoul, 17–20 Oct 1994, pp 890–895
9. Jeong S, Choi J, Kim J, Lee Y, El-Keib AA, Shahidehpour M (2008) Flexible best generation mix for Korea power system considering CO_2 Constraint—Vision 2030, IEEE PES GM2008, 20–24 July 2008, Pittsburg, PA, USA
10. Park J, Wu L, Choi J, Cha J, El-Keib AA, Watada W (2009) Fuzzy theory-based best generation mix considering renewable energy generators. Fuzzy-IEEE2009, 20–24 Aug 2009, ICC Jeju, Jeju Island, Korea
11. Goldberg DE (1989) Genetic algorithms in search optimization and machine learning. Addison-Wesley, Reading, MA
12. Levin N, Zahavi J (1984) Optimal mix algorithm with existing units. IEEE Trans PAS-103:954–962
13. Levin N, Zahavi J (1985) Optimal mix algorithm with limited-energy plants. IEEE Trans PAS-104:1131–1139
14. Dhar SB (1979) Power system long-range decision analysis under fuzzy environment. IEEE Trans PAS PAS-98(2)
15. Bloom (1982) Long-range generation planning using decomposition and probabilistic simulation. IEEE Trans PAS 101:4:797–802
16. Sullivan RE (1977) Power system planning, McGraw-Hill
17. Billinton R (1970) Power system reliability evaluation. Gordon and Breach, New York
18. Billinton R et al (1973) Power system Reliability Calculations, MIT
19. Yasuda K et al. (1988) Optimal generation expansion planning with electric energy storage systems: IECON, pp 550–555
20. Takahashi K et al (1988) Power systems flexibility, principles and means, available methods at the planning stage. CIGRE SC-37, Brussels, Feb 1988
21. Arakawa F et al (1989) Energy security assessment quantitative analysis for flexible system planning. CIGRE SC-37, Florence, Oct 1989
22. Tanabe R et al (1992) Determination of robust generation mix considering uncertainties. JIEE Jan 112-B(4):331–338
23. Jenkins N, Allan R, Crossley P, Kirschen D, Strbac G (2000) Embedded Generation, pp 31–38 (2000)
24. Divan D, Luckjiff DA, Freeborg JW, Brumsickle W, Bielinski D, Grant T (2000) I-Grid™: a new paradigm in distribution grid power quality and reliability monitoring. In: Proceedings of EPRI PQA north America conference. Portland, OR (2002)
25. Billinton R, Gao Y (2008) Multi-state wind energy conversion system models for adequacy assessment of generating systems incorporating wind energy. IEEE Trans Energy Convers 23 (2):163–170
26. Park J, Wu L, Choi J, El-Keib AA, Watada J (2009) Probabilistic production cost credit evaluation of wind turbine generators. Int J Innovative Comput Inf Control (IJICIC) 5(11 (A)):3637–3646
27. Billinton R, Huang D (2008) Aleatory and epistemic uncertainty considerations in power system reliability evaluation. PMAPS'08, Puerto Rico, 25–29 May 2008
28. Karki R, Billinton PHR (2006) A simplified wind power generation model for reliability evaluation. IEEE Trans Energy Convers 21(2)

29. Choi J, Kim H, Cha J, Billinton R (2001) Nodal probabilistic congestion and reliability evaluation of a transmission system under deregulated electricity market. In: IEEE, PES, GM2001. Vancouver, Canada
30. Choi J, Billinton R, Futuhi-Firuzabed M (2005) Development of a new nodal effective load model considering of transmission system element unavailabilities. IEE Proc G,T&D 152 (1):79–89
31. Wu L, Park J, Choi J El-Keib AA (2009) Probabilistic reliability evaluation of power systems including wind turbine generators considering wind speed correlation, KIEE. J Electr Eng Technol 4(4):514–519
32. Wu L, Park J, Choi J, El-Keib AA, Shahidehpour M, Billinton RR (2009) Probabilistic reliability evaluation of power systems including wind turbine generators using a simplified multi-state model: a case study. IEEE PES GM, Calgary
33. Park J, Cho K, Oh T, Choi J, El-Keib AA (2010) Evaluation of economic impacts of wind turbine and solar cell generators on power system considering CO_2 Emission. PMAPS2010, Grand Copthorne Waterfront Hotel, Singapore
34. Billinton R, Li W (1994) Reliability assessment of electric power systems using monte carlo methods, p 24–30
35. Choi J (1990) An efficient algorithm for probabilistic operating simulation. Ph.D. thesis, Aug 1990

Chapter 8
Microhydro-Based Mini grid for Sustainable Development of Rural Communities: A Case Study of Nepal

Netra Pd. Gyawali

Abstract This chapter titled "Microhydro-Based Mini Grid for Sustainable Development of Rural Communities: A Case Study of Nepal" presents the overview of Microhydro-Based Mini Grid for Sustainable Development of Rural Communities in Nepal. Development of energy sector in Nepal demands optimum tapping of hydropower of different scales; be it through large, medium, or small size hydro plants. Large power plants are often required big capital investment and involve long constructional time, whereas small and microhydro plants (SMHP) can be installed in short time period with local initiative. Since very beginning SMHP are proved to be as cost effective, indigenous, and reliable source of electrical power. However, they also have been facing hurdles in full scale development. The main issues are low load factor, low capacity to run big machines, poor end-use pattern, and unavailability of water throughout the year. Besides the expansion of national grid to the SMHP vicinity has caused to extinct, thereby wasting the productive asset. This scenario has forced to develop integrated hydro power development model in country, in which SMHP can be operated even after the reach of grid. By this, additional revenue will be generated from surplus electricity. Second, parallel operation of two or more SMHPs can realized for sharing the common loads. Formation of such MGs would have number of benefits; it would maximize the use of spilling resource, increase load factor and improve the reliability of the overall system. These MGs can be connected with national grid if it is nearby, and operated autonomously in islanded mode and interconnected mode in specified manner. From utility perspective, the MG functions as a single entity, which can provide a better platform to interconnect SMHPs to grid. Thereby it can be operated even after the reach of grid in their sites, making MG a sustainable scheme for SMHP development. There are several sites in Nepal with potential of MG. Alternative Energy Promotion Center (AEPC) has reported 14 sites across the country with capacity ranging from 100 kWs to 1 MWs. AEPC also formed a pilot MG project in Baglung District by connecting the SMHPs along the corridor of Kalung Khola river, which was completed and commissioned in 2011. Presently,

N.Pd. Gyawali (✉)
Department of Electrical Engineering, IOE, Tribhuvan University, Pulchowk Lalitpur, Nepal
e-mail: netra@ioe.edu.np

N.R. Karki et al. (eds.), *Sustainable Power Systems*, Reliable and Sustainable Electric Power and Energy Systems Management,
DOI 10.1007/978-981-10-2230-2_8

It has been drawing attention to the researchers, entrepreneurs, and professionals across the globe for the real-time study of MG. Its success stories have paved the ways for the further development of other MGs in the different parts of country.

1 Introduction

1.1 Microhydro in Nepal

Nepal is a third world country sharing common features of developing nations, such as low energy per capita consumption and dominancy of firewood as the fuel source, among others. These are typical indicators rendering developing countries be more vulnerable to climate change. Looking into the entire scenario of energy use of the country, the electricity is a tiny fraction, in which only about 5 % energy need is fulfilled by electricity. The bulk of the energy supply is dominated by fuel wood (68 %), agricultural waste (15 %), animal dung (8 %), and imported fossil fuel (8 %) [1]. Only about 60 % of Nepal's population, mainly living in urban/suburban area, has access to electricity. Having immense potential of hydropower, it is important for Nepal to generate maximum hydroelectricity and increase its energy dependency on electricity. This not only contributes to protect environment, but also to mitigate deforestation and soil erosion, increasing literacy, and improving health of children and women with green energy use.

Nepal electricity authority (NEA), the government-owned utility, is the main stakeholder for the electric power sector development in Nepal. It has dominated the power generation system, with little share of independent power producers (IPPs). At present, total installed power generating capacity is 720 MW; of which, 88 % is supplied by hydro-based generation. With current total demand of 1100 MW, the available generation capacity falls way short and the country is facing severe power deficit with long hours of load shedding during most of the time in a year. This situation has drawn the compulsive attention of deploying as much new generators as we can; be it through large, medium or small scale hydro plants, and other renewables. Large power plants are often required big capital investment, likely through foreign companies, and involve long constructional time, whereas small and microhydro plants (SMHP) can be installed in short time period with local initiative and operated by local manpower, and hence proved to be sustainable source of electricity for rural community.

Development of SMHPs for electrification in Nepal was started in 1960s by the initiation of some enthusiastic entrepreneurs to electrify a rural community of Kathmandu Valley. It was the milestone in SMHP developments in which the traditional SMHPs aimed for agroprocessing were evolved to generate electricity as an add-on function. Following the installation of the first SMHP, there was widespread deployment of SMHP throughout the country, resulting substantial change in socioeconomic activities of rural people. The government of Nepal, then, realized to form a formal body to streamline SMHP, and consequently alternative energy promotion center (AEPC) was established in 1996. To accelerate the SMHP deployment in remote area, AEPC formulated a subsidy policy, in which 75 % of installation cost would be borne by the state. The subsidy is provided as a part of promotion program that links financial support with a structured project cycle, productive end-use, and quality requirements. The subsidy policy contains the following key provisions: [2].

- NRs 55,000 per kW for new microhydro projects up to 3 kW capacity (mainly peltric sets)
- NRs 70,000 per kW for new microhydro projects above 3 KW
- NRs 35,000 per kW or 50 % of costs for rehabilitation projects
- From NRs 8750–21,000 per KW as a transport subsidy
- NRs 27,000 for add-on electricity generation from improved Ghattas
- An investment ceiling per kW of NRs 150,000

As a result, AEPC could able to install 2500 SMHPs in hilly and mountains locations, amounting to 20,000 kW of installed capacity, which has been supplying electricity for about 200,000 households. It is planning to add another 10 MW of SMHP in coming five years electrifying several thousand households.

1.2 Why Mini Grid?

While Micro hydropower projects have gained popularity as a cost effective, indigenous, and reliable source of electrical power, they have also been facing hurdles in full scale development. Installed in the isolated areas, SMHPs are mainly supplying the lighting needs of the local community. By virtue of poor load factor of the lighting loads, their productive end-uses is limited. Even though multiple end-uses are possible to increase the utilization factor, it is somehow impossible to achieve the optimum capacity factor due to constraints associated with a single plant. Also, there is difficulty with SMHP in providing round the clock supply for plants that share water for other important purpose like irrigation. Further, the flow of small streams may restrict future site expansion as the power demand increases. In many locations, stream flow rate will fluctuate seasonally. During the dry months there is less flow and therefore less power output.

Besides, there is expanding of national grid and it is reaching to the SMHP vicinity, causing SMHP consumers to switch toward the grid supply. It has been reported that grid has already reached to 11 % of SMHP sites, and it will expand to another 27 % of SMHP sites by next 5 years. Due to the lack of proper interconnection plan, the utility is reluctant to accommodate small scale generators. The major concern is that grid has to entertain large number of SMHPs individually, which will aggravate the safety issues during the repair and maintenance of grid work. According to Electricity Act (1992), NEA is obligated to purchase the micro-SMHPs from the owner if the grid penetrates the SMHP areas. However, the management of small plant is cumbersome for NEA and practice has been to abandon the plant after purchasing. As such, many SMHPs were already discarded and others are likely to be so in incoming years. Thus the expansion of the grid has posed a threat for continue the operation existing SMHPs, thereby wasting the productive asset and also hindering the development of new micro hydro projects.

This scenario has forced to develop integrated hydropower development model in country, in which continued operation of SMHP should be ensured. Thereby additional revenue will be generated to the community by selling surplus electricity to grid, which otherwise was supposed to discard. In addition, maximum utilization of installed SMHP have direct influence in reduction of greenhouse gases enhancing the sustainable development of rural community. Second, the technology of Mini grid (MG) could be adapted, in which two or more SMHPs can be operated in parallel forming a miniature grid. MG is an emerging technology, in which small generators at low voltage electrical network are designed to supply electrical loads for a small community, such as remote villages, commercial areas, industrial sites, etc. Such an MG can be operated in autonomous manner both in islanded and grid connected modes, which would provide several added benefits. The load factor of the system will increase as the surplus power of one area can be transferred to the power deficit area. The service reliability will also be improved as the local consumers could get supply even though the respective SMHP is down. From grid perspective, the MG would provide a better platform to its interconnection with national grid; where the national grid can treat the mini grid as a single entity, instead of dealing with several small scale SMHPs as “negative load.” Thereby it can be operated even after the reach of distribution network in their sites, making MG a sustainable scheme for SMHP development.

There are several sites in Nepal with potential of MG. AEPC has reported 14 sites across the country with capacity ranging from 100 kWs to 1 MWs. It also formed a pilot MG project in Baglung District by connecting the SMHPs along the corridor of Kalung Khola river, which was completed and commissioned in 2011. Presently, It has been drawing attention to the researchers, entrepreneurs, and professionals across the globe for the real-time study of MG. Its success stories have paved the ways for the further development of other MGs in the different parts of country.

2 Mini Grid Layout

2.1 System Overview

The MG under consideration, also known as Baglung MG, is situated at the south-eastern part of Baglung district along the corridor of Kalung Khola river, known as *Urja Valley*. Seven SMHPs were installed from the river in the adjoining village development committees (VDCs), namely Sarkuwa, Rangkhani, Damek and Paiyu Thanthap (Fig. 1). The SMHPs with capacity ranging from 9 to 26 kW are connected to 11 kV network through respective 0.4/11 kV transformers and

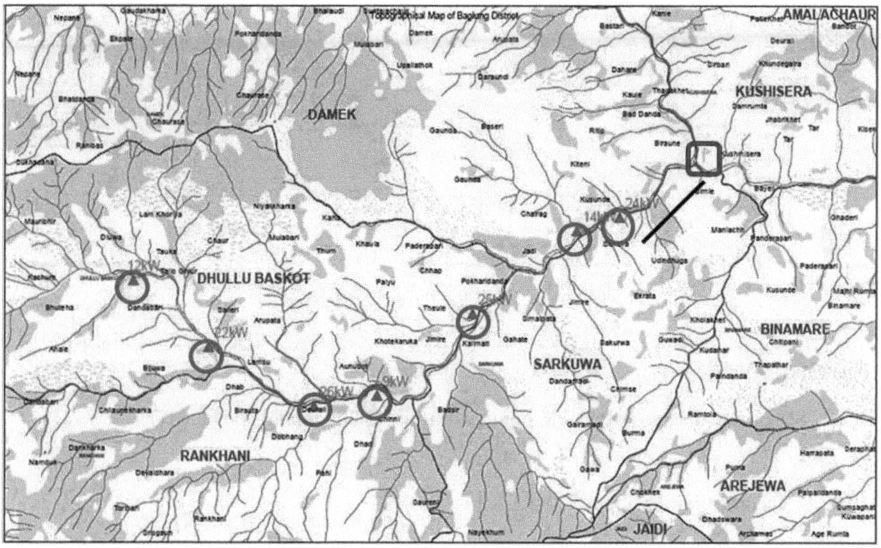

Fig. 1 Topographic map showing six SMHPs forming the mini grid [3]

synchronized by automatic synchronizer. At present net generation capacity is about 132 kW and serving about 1400 households of the locality. The average load factor of the mini grid is about 0.47 with its peak load of 127 kW at the evening, implying that there is surplus of power at the off peak time. Starting from the far end of SMHP located at Bijuwa (Paiyu) to Theule Khola (Sarkuwa), the total length of 11 kV network is about eight kilometer. As shown, the MG is only 1 km far from the existing 33 kV line, implying that there is a possibility of connecting the MG with the NEA grid.

The single line diagram of the MG is presented in Fig. 2. Each SMHP consists of a synchronous generator set, 0.4/11 kV transformer, an ELC (electronic load controller)-based ballast load and a synchronizing unit. The power is generated at 0.4 kV, which then supplied to the local load and also transmitted to 11 kV through the stepup transformer. Since the generated power is not controlled, the ballast load consumes the surplus power to balance generation and demand, and consequently the system frequency is regulated at designated value. Likewise, the synchronizing unit facilitates the interconnection of one generator with the others generators, thus enabling to form and expand MG system.

Typical single line diagram of a SMHP unit is shown in Fig. 10 (Appendix). The generating system is equipped with numbers of control functions, protecting relays, metering units, and switching elements. The SMHP control system consists of two functions; the automatic voltage regulator (AVR) and ELC. The AVR and ELC take care of excitation control and active power balance, respectively; both in isolated and interconnected modes through droop mechanism. There are number of relays associated with the generator protection schemes, namely over current (O/C)

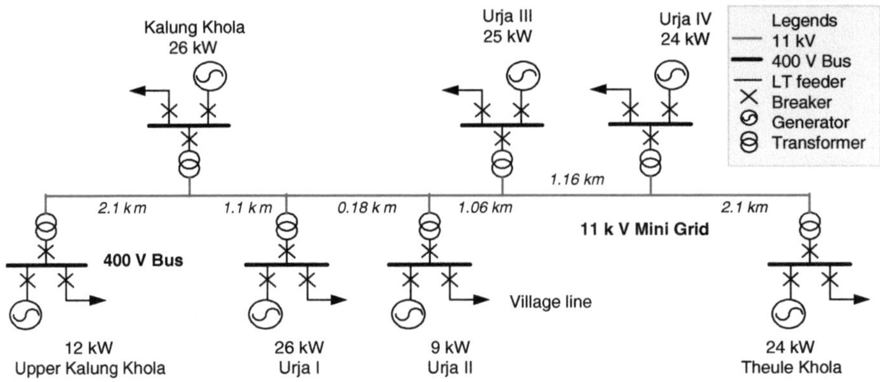

Fig. 2 Schematic diagram of 11 kV mini grid in the Kalungkhola corridor [3]

Table 1 Specification of upper Kalung Khola SMHP

S. No.	Item	Detail
1	Generator Rating	50 kVA
2	Excitation	AVR with droop facility
3	Protection	Under/over voltage Reverse power relay Earth fault relay Voltage restraint over current relay Under/over frequency relay
4	ELC ballast load	24 kW full load, grid synchronizable, thyristor controlled
5	Transformer	11/0.4 kV, 40 kVA
6	Metering	Voltmeter with selection switch Ampere meter selection switch kW, kWh meter Power factor meter Frequency meter, fault announcer

relay, over and under voltage relay, reverse power relay, and so on. Further, the power plant also has two energy meters, one for measuring the power generated by the SMHP and the other for measuring net power consumed by the local load. The details of these functionalities are also depicted in Table 1.

2.2 *Details of System Components*

2.2.1 ELC System

Electronic load controller is a device that maintains constant load to the generator despite the variation of the load in the village. This is done by diverting surplus

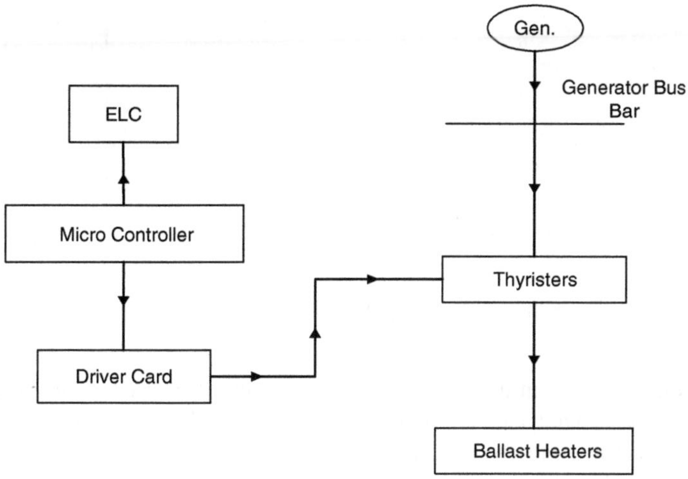

Fig. 3 Layout of ELC system

power to the ballast load, keeping the frequency at designated value despite load change. The schematic diagram of ELC applied to system is shown in Fig. 3. The ELC unit consists of ELC block, gate driving unit and ELM. The input signals to ELC are generator voltage, grid voltage and signals from the relaying units. The input signals are processed by the ELC unit and sent to driving unit (Drive Card) and ELM unit. The main function of the drive card is to generate the gate signals to thyristor as function of the firing angle. It also helps to suppress the spikes while chopping the ac cycles. Auxiliary parts of the ELC block are voltage transformers and dc power supply. The ELC system functions differently in isolated and grid connected modes. In isolated mode, ELC operates so as to make the system frequency at the constant value, while at the interconnected mode, power sharing takes place through droop mechanism, thereby frequency is allowed to vary with in certain limit (2 % in this case). That means the frequency drops from the rated value (no-load state) to 98 % at its full load state. Another importance of frequency droop is that there is proportional sharing of added load to the system. Suppose two generator having capacity 50 and 20 kVA having droop setting of 2 %, both of them operating at no-load. If a load of 20 kW is switch on then, first generator will share 14.3 kW and second generator will share 5.7 kW of load (Fig. 4a).

2.2.2 AVR Unit

The main function of the AVR is to regulate the generator voltage at designated value despite of variation of load. AVR operates in two different modes; isolated and the MG modes. In isolated mode, it regulates the terminal voltage at the constant value, while in the MG connected mode, voltage regulation is performed

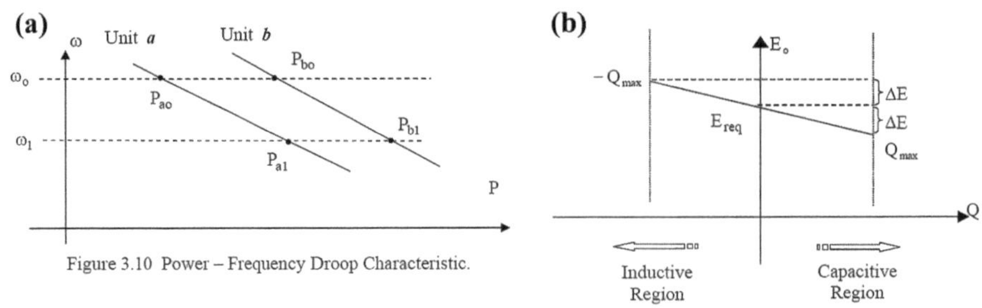

Fig. 4 Droop scheme for MG; **a** *P-x* droop, **b** *V-Q* droop

by droop regulation technique. AVR with droop voltage characteristics helps for the parallel operation. That means when load increases AVR must decrease its excitation, thereby lowering. In this system 4 % *V-Q* droop is set. For example, if no-load voltage is set at 400 V, the full loaded terminal voltage will be 384 V, while proportionally sharing the applied reactive load. Suppose two generator having capacity 50 and 20 KVA having droop setting of 4 %, both of them operating at no-load. If a load of 20 kVAR is switch on, 50 kVA generator will share 14.3 kVAR load, and the 20 kVA generator will share 5.7 kVAR load (Fig. 4b).

2.2.3 Switchgear and Protection System

Protection system is the primary requirement of the power system. For the protection of power system equipment in abnormal situation, protective equipments are provided in control panel, such as fuses, MCBs, MCCBs, contactors, relaying system, and emergency switches. The features of the switching elements are summarized in the following paragraphs.

- *Fuses:* Fuses are much conventional type of protection elements. They are placed in series with the element that has to be protected. In this case small fuses are provided in all equipment, such as AVR, ELC, diver card, battery charging units, etc.
- *MCBs: MCBs* are superior protection element than fuse. They can be operated manually or automatically during over current. The control panel uses eight MCBs to provide protection from over current to the different circuit branches.
- *MCCBs:* Two MCCBs have been used in our control panel, one MCCB provides switch for village side and the other provides for grid side. They can be operated manually or automatically by providing 24 V supplies for to its shunt trip coil. Both village and grid MCCB are operated by the action of master trip relay depending upon the nature of fault. Both MCCBs can also be operated manually.

- *Contactors:* In case of contactors, DC and AC contactors are used in the control panel. The DC contactor requires 24 V dc supply to operate and it provides switch for DC system, whereas ac contactor requires 220 V ac supply to energize. It provides protection to the generator and blocks the generator supply.
- *Relays:* The Relaying system consists of number sensors to sense abnormal condition in the circuit and giving the tripping signal for the MCCB and Contactor. These relays operate in three modes; scan mode (gets activated for searching of faults), set mode (setting the value of tripping parameters), and value mode (shows the instantaneous values of the parameters). The main relaying functions include:

(a) Over-under frequency relay	(e) Voltage restrained over current relay
(b) Over-under voltage relay	(f) Reverse power relay
(c) Over current rely	(g) Earth fault relay
(d) Phase unbalance relay	(h) Master trip relay

The objective of these relays is to protect the generator and LV feeder from abnormal condition. The relays give the signal for the Master trip relay which will trip the AC contactor and village MCCB. There are six auxiliary relays used in the control panel which provides both AC and DC supply for various equipment at suitable instant and are energized by 24 V DC supply (Fig. 5).

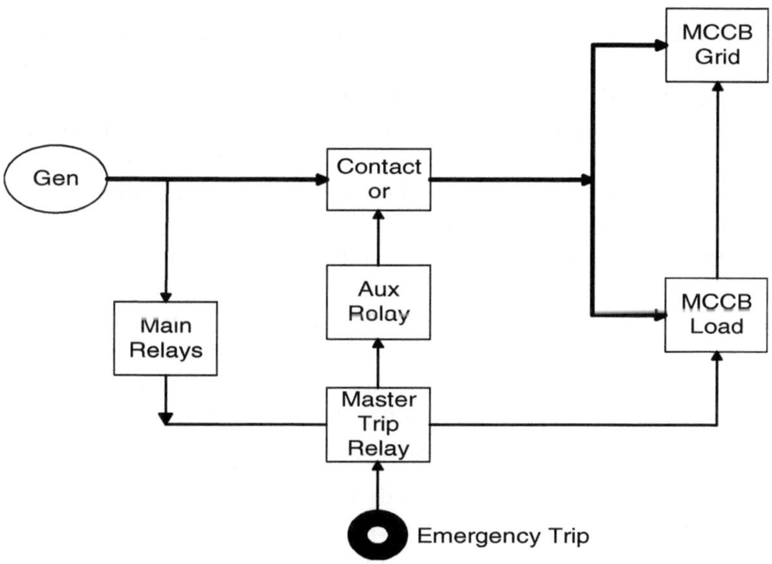

Fig. 5 Layout of protection and relay system

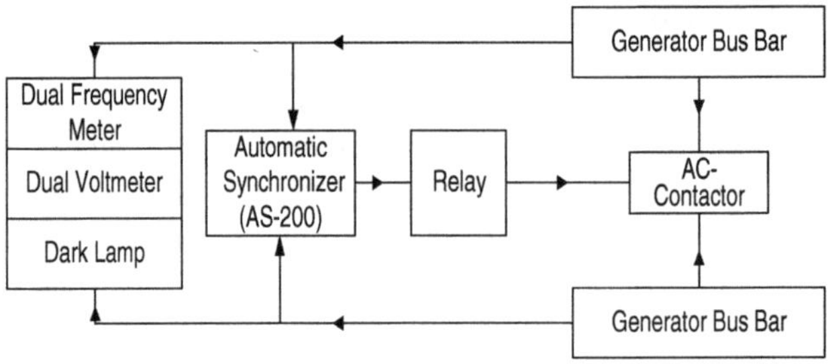

Fig. 6 Synchronization scheme

- *Emergency tripping unit* It is the most important unit in the control panel. It allows the operator to disconnect the generator and village load from grid if desired. If the operator feels any abnormal condition inside power house or village distribution system or grid transmission system, he can operate emergency key which prevent the generator to supply to village and grid. It also trips the LV feeder so the local load can be disconnected.

2.2.4 Synchronization

Synchronization is a process of operating generators or a generator with grid in parallel. For the synchronization three parameters—*frequency, voltage,* and *phase sequence*—should be matched. Frequency control is achieved through ELC, while voltage is controlled through excitation. The synchronization can be carried out either by automatic or manual mode. In auto mode, contactor is automatically closed and a connecting key is used in manual mode. The block diagram of synchronization is shown below (Fig. 6). During auto synchronization process, when synchronization enable switch is turn on, auxiliary relay gets their ground supply, gets energized, and closes path for generator and grid signal to dual meters.

During synchronization process, dual voltmeter, dual frequency meter, and auto synchronizer provide this comparing system. Dual voltmeter shows the grid and generator line to line voltage and becomes enabled only synchronization process. Also, dual frequency meter shows grid and generator frequency during synchronization; whereas in isolated mode, it only shows generator frequency. Automatic synchronizer shows information on whether the

- grid and generator voltages are exactly equal,
- grid and generator frequency are exactly equal,
- phase sequence of both voltages is exactly matched

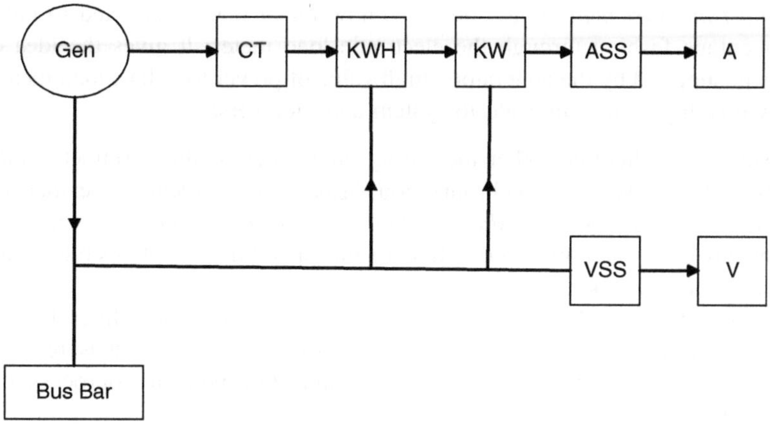

Fig. 7 Layout of metering scheme of SMHPs

2.2.5 Measurement System

Measurement of system variables is very important since it provides the idea of what is going on inside the system and what sort of actions need to be applied for corrective action, if any. The measurement system monitors the variables from the different part of the system, as detailed below.

- *Generation measurement system* includes monitoring values of voltage at generator output terminal and output current injected by generator after the ballast load (ballast load is not included). For current measurement, three current transformers are installed in between generator and contactor in vertical bus bar while for voltage is directly measured through PT. The block diagram of metering system adopted for the generation measurement is shown below in the block diagram (Fig. 6 in Chap. 5).
- In *Village load measurement unit*, current supplied to the local load is measured by three current transformers after ac contactor in village bus bar. Ammeter shows phases currents and voltmeter shows line voltage as selected by the respective selection switches. Village load measurement system shows information about total kW being consumed, total kWh consumed, power factor of load, line voltage, and the line current (Fig. 7).
- *Excitation unit* of the SMHP is comprised in AVR block, for which an ammeter and a voltmeter is provided in the control panel for acquiring the current and voltage data. Ammeter shows the magnitude of excitation current, whereas voltmeter shows the magnitude of excitation voltage. These values should not exceed the present values provided on the generator name plate otherwise generator may be overheated and may cause undesired damages.
- *Battery voltage measurement system* measures the battery voltage with a DC voltmeter (provided in the control panel), showing the current.

- *Ballast measurement system* observes how much load is diverted to ballast in percentage fashion through the electronic load meter. It gives the idea of net power supplied by the generators; high value of power to ballast load means that low quantity of it is injected to system and vice versa.

Beside these, there are other measuring units; such as dual voltmeter and dual frequency meter used for synchronization, *Semaphore* to identify weather the AC contactor is on or not, and *Alarm annunciator/electronic hooter* to provide the information of the abnormal condition to the operator so that operator can take necessary corrective action.

To sum up, each SMHP is equipped with necessary technicalities for parallel operation. In earlier system the option was not available, which is replaced by aforementioned tools and schemes for grid connection with microprocessor-based control panel.

2.3 Management Model

Since the MG is the expanded form of SMHPs, both in size and attributes, the management of the MG is quite complex than the standalone systems. Two separate entities are involved in the MG; namely SMHP operator and mini grid system operator (MGSO). The SMHP operator lies at the bottom level and responsible for the generation of electricity, operation, and maintenance of plant, functioning as independent power producer (IPP), whereas MGSO lies at the top level and responsible for transmission, distribution, operation, and management of MG at the 11 kV and 400 V level. The MGSO also sets electricity tariff for households based on the mutually consent. In this case, buying rate from IPPs is NR 4.5 per unit and selling rate to consumers in NR 5.75/unit for initial 12 units (minimum) and NR 7/unit then onwards.

Besides, MGSO follows the role and the regulation of cooperatives. The article and memorandum of understanding, financial and human resource plan and regulation, formulation of load management, and systems maintenance and the business plan are prepared to run the cooperative smoothly. The MGSO functions through *Mini Grid Working Committee* registered *Urja Upatyaka Mini Grid Cooperative Ltd* in September 2011. All 1200 HHs who are the consumers of the SMHPs are members of the cooperative. The Executive Committee has been formed with representatives from the consumers. Three professional staff, two with technical background and one administrative, have been assigned to run the project smoothly.

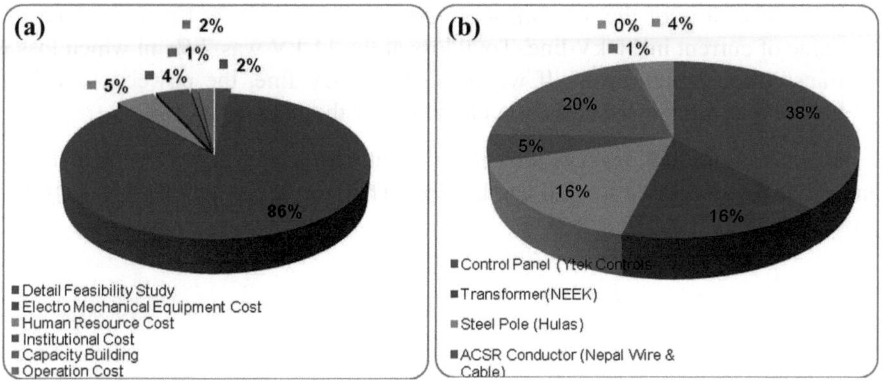

Fig. 8 Cost Breakdown of Baglung mini grid **a** Total project cost, **b** Electromechanical cost (ReRL report)

2.4 *Financial Model*

The project is financed by UNDP/REDP and the community is actively involved in construction and operation of the project. The project cost of MG is 14.5 Million, with dominance of equipment (86 %), in which control panel allocates 36 % of equipment cost. The financial support is still needed for institutionalization of the cooperative, smooth operation, repair and maintenance, and capacity building of the people involved in mini grid. The overall cost (SMHPs plus MG) reaches to 0.32 Million/kW, which is costlier than normal SMHP system. However, the cost can be lowered significantly, if the technology can be realized in domestic level (Fig 8).

2.5 *Network Analysis*

In order to estimate, the loss, voltage profile, fault level, and power flow through branches, a detailed load flow analysis was carried out of the MG. The MG is radial in nature with conductor having high value of resistance. For this, *WindMil*^{TM} software was used and the results are depicted in Fig. 12. The salient features of the results are summarized as

- Load profile is smooth with no swell and sags in any branches. It is because the entire generators are regulating their voltage with *V-Q* droop, and there is no circulating VAR among the generators. This justifies the comfortable level of voltage regulation in the network.

- The power loss of the network was very small below 2 %, because of the low value of current in 11 kV line. Total loss at the 11 kV was 7 %, in which loss in transformer is significant. If we consider the LV line, the power loss will be higher than it. Nevertheless, the efficiency of the network is fine.

The maximum fault level of the network is at with approximated value of 1.6 kA and gradually decreased toward another end. The fault levels guide the selection of protecting relays and switch gear.

3 Impacts

After the commissioning of the MG, the localities have observed number of benefits in different aspects. The consumers and the IPPs both have gained technical, social, and financial benefits by its operation. Following subsection summarized the benefits observed by different disciplines.

3.1 Technical

After formation of MG, the localities have observed following impact:

- There is technical upgrade in SMHPs after the introduction of MG Technology, because the older control technologies were compatible with MG operation. These technology have resulted stable voltage and frequency, provide more reliable service. Further, fault detection process is made easier and faster by using relays-based protection.
- The quality of electricity has been dramatically improved; the voltage level of the distribution is regulated within 10 %, paving the way for motor drive applications.
- Overload in a generator has been decreased due to load sharing, making load plant higher than previous value.
- Mini grid is becoming an observatory place for national and international researchers. Different organizations in energy field are interested to replicate the technology and management model in other place as well.
- MG is converting as an entity for generation, transmission, and distribution of rural energy. It also assures the space for connecting small SMHPs, which otherwise was not allowed by NEA.
- The large scale end-use application is possible with availability of large surplus power. For example, NCell mobile communication system is powered by MG; 40 HP stone crushing machine is planned to install and run during the off-load period

3.2 Social

The social benefits of MG use are summarized as follows.

- Mini grid is becoming the symbol of unity that unites the seven communities not technically but socially as well. This motivates continued operation of SMHPs even after the reach of grid to the locality. Every individual in the community owns the project and participate in it activities
- Misuse of electricity and electricity theft in the community has been decreased, while productive end-use application is heightened.
- Entrepreneurship skill of community is developed. They are willing to construct, own, and manage bigger projects.
- It has been easier for the educational institution to provide IT education and facility to the students. Communication, entertainment, and news listening habitual change is seen in the rural people due to availability of energy at day time for households. Around 10 kW HHs load increased due to television, computer, and other electronic loads
- Inter community coordination seen due to the establishment of Cooperative in the mini grid society. Seven different community groups merge into one huge community and became members of common institution.

3.3 Financial

The local people are financially benefitted in following ways.

- Income of power houses has increased. Basically power houses of capacity greater than 15 kW has increased their income around NPR 5000–6000 while power houses with capacity less than 15 kW increased their income around NRs 2500–3000. Further possibilities of financial strengthening of the SMHP and MG are seen. End use like N cell tower demanding 10 kW has been possible and there is plan of installing big stone crushing plant. They would contribute in financial sustainability of the project.
- It creates more than 55 job opportunity for local people through the establishment of Agroprocessing mills, Poultry farming, Computer Institute, Photo studio, electronic shop, etc. Also, the income of entrepreneurs has increased as they can be used electricity any time of day and night as their requirement.

In short, the local people have been substantially benefited from the MG, in term of technical, financial, and social aspects. Some advantages are tangible and can be measured directly in term of monetary unit, other are indirect having long-term importance on the society.

4 Grid Connections

Formation MG also provides better flexibility of integrating SMHPs to the utility grid. Instead of dealing with large number of SMHPs, utility would deal with a single entity. Thus MG would render itself a sizable load (generating unit) from utility perspective. This will encourage small SMHP owner to join into electricity through MG, thereby maximizing the revenue and improving the supply reliability. Alternately, interconnection is basically meant for back wheeling the excess power produced by the MG to grid, and if necessary MG can also purchase power from grid during the time of power deficit. To tap these benefits, RERL/AEPC is planning to connect the MG with grid as a pilot project. The detailed feasibility study has already been carried out, suggesting that the project is highly feasible. However, there are some technical and operational issues to be addressed for maximizing the benefits. These issues are summarized below.

4.1 Issues

4.1.1 Grid Side Concerns

From utility perspective, connection of large number of small SMHP at the distribution line may cause additional protective schemes to be implemented. The first is associated with the formation of islanded power network. This is related to safety of the persons who are working on the electrical system. When power plants are connected to the radial feeder of a distribution network, the feeder becomes an active network and the back fed takes place in the islanded condition. There is potential risk to the maintenance crew if the MG side is still feeding the network. The second is isolation of MG with grid during the fault inside the MG and provide necessary protection to power network: This emphasizes the timely disconnection of MG breaker before the breaker at the substation trips, otherwise the fault inside the MG will increase the outage time of the distribution network [4].

In short, the interconnection demands a suitable anti-islanding protection scheme and a switching coordination between the MG breaker and substation breakers. There are various islanding detection schemes detection techniques, from highly sophisticated (active detection techniques) to the simple one (passive detection techniques). Among the passive scheme, combined *rate of change of frequency (ROCOF) relay* and *voltage vector shift (VVS) relay* is widely used with minimum non-detecting zone during the internal power balance situation. The VVS/ROCOF protection relay is designed for applications where a generator is running parallel with main supply either from the utility or other generators, in which the relay detects disconnection of a generator from the network. The VVS relay measures the length of each cycle of the voltage wave. At the moment the grid becomes disconnected, the sudden change in load causes a sudden change in cycle length, causing an abnormal voltage phase shift. Similarly, the frequency of island will change rapidly due to active power unbalance, and the corresponding frequency slope can be used to detect loss of main.

4.1.2 Mini Grid Side Concerns

When mini grid is operating in grid interconnected mode, with the assumption that the grid is too strong, the grid frequency and voltage dictate those of mini grid. Under that condition, mini grid should operate at active and reactive power control modes, in which the ELC ballast load should draw no power. Critical situations can occur if the upstream power network is disconnected due to a fault and the mini grid is islanded. This situation is commonly referred as islanding. When the islanding occurs, there should be an autonomous voltage and frequency control at the individual SMHPs. In such case, well-established technology is to employ droop control techniques in both active and reactive power regulations. By applying the droop method, the excess active power and reactive power would be shared, respectively, by the ballast loads and the generators, in proportional to their rating. The existing SMHPs are already equipped with droop regulation scheme to control active and reactive power, which enables operating in islanded and grid connected modes smoothly. However, for grid interconnection, a relaying scheme is needed that quickly identifies the islanded state and disconnects the breaker at the connection point. The islanding detection scheme described earlier server the purpose.

4.2 *Proposed 33/11 kV Substation*

Since the existing NEA feeder is of 33 kV, there is necessity of 33/11 kV substation with MG for interconnection. For this, 33/11 kV, 250 kVA substation is proposed at Sarkuwa VDC. The substation includes the necessary safety/protection measure and the metering facilities. It should, however, be considered that if existing 33 kV line would be charged at 11 kV after few years, it is desirable to interconnect the mini grid at 11 kV level. The 11 kV interconnection will be simpler and cost effective when compared to 33 kV interconnection [5].

The proposed single line diagram of 33/11 kV switching station is shown in Fig. 12, presenting necessary protective and synchronizing relays, switch gears, and metering system. In this case, *ROCOF* and the *VVS* relay are used for islanding protection.

5 Prevailing Issues and Wayforwards

Construction of MG with local initiatives was a big challenge in developing countries. Though it is addressed and the project completed successfully, there are few technical and managerial issues which may affect sustainable operation of the MG.

- *System Management* The management of MG system is more complicated than the existing isolated system. It involves the people with diverse interests. Therefore, good coordination among IPPs and MGSO is necessary to follow the power purchase agreement, also the MGSO should render itself financially sound to meet the operational and maintenance cost of the system.
- *Human resource for operation and maintenance* Being an imported technology, local repairing and maintenance service is not difficult. Therefore, the technology has to be internalized within the country to avoid delays in procuring and maintenance services. Further, the technology transfer and the capacity building to the local people is desirable.
- *Load management and end-uses scheme* Presently, the load factor is low with much space for *installing* new and high-power consuming equipment. This deserves special effort to identify and promote additional end-uses. Likewise, load management in the grid an important feature. The system is connected through the MG but there is no central unit which can monitor and control the performance of the MG components automatic manner. The load management plan has been prepared for 24 h with mutual understanding among the SMHPs, load management.
- *Grid operation in dry season* The demand in the mini grid system is 100 kW at peak hours and the supply is 107 kW. However, in dry season the supply will be less due to the use of water for irrigation purpose that may affect the supply systems.
- *Technology transfer* since the technology was imported, it is necessary to internalize the technology and training local people for operation and maintenance.

Being a pilot project, there is always risk factor of the desired operation, and the sustainability of such project is not very clear in advance. Therefore many activities are introduced to making MG sustainable. The key activities and efforts involve:

- *Capacity building* AEPC/RERL has supported and made strategy to build the capacity of the different stakeholders in mini grid at different level. At the local level, training to operators during and post-installation, training to staff of the cooperative and orientation to executive members of cooperative for better management of cooperative. At the central level, AEPC has initiated

collaboration among academician, government officials and entrepreneurs for their participation in sustainable development of SMHP and MG technology. This includes incorporating research activities in academic institutes, manufacturing control panels by local industries, and so on.
- *Support for possible end-uses* Special efforts have been given to promote potential end-uses in MG area. Various trainings have provided to local community in order to motivate them toward income generating activities through the uses of locally available resources. Livelihoods Promotion Officer has already focused in that area and a Chilling Vat & a Cold store are being installed in near future.
- *Monitoring and evaluation* Being the pilot project, the projects need close monitoring in terms of technology and management. Support from AEPC needs to be continued for at least one more year for the effective and sustainable operation of the project.
- *Initiation for grid connection* Grid connection initiation has moved ahead. Grid connection will benefit the community by opening new avenues for generating additional income by selling electricity to the NEA. Special effort is needed to materialize the grid connection of mini grid as a pilot project.

6 Conclusions

The development of MG is growing stage in Nepal with one project which has been installed as a pilot one. This project has resulted many positive impact to the localities and paved the way for deployment of MGs in scale through the country. The outcome has inferred SMHP-based MG scheme to be a cost-effective means of sustainable and reliable power to rural communities. The key finding of the study has been pointed out as follows.

- Micro hydro technology, being important source of electricity, should be developed and deployed extensively wherever possible. Formation of MG is an effective way to leverage optimum development of SMHP, in which a cluster SMHPs can be operated as a single entity in an autonomous way. Such a mini grid would significantly enhance quality, availability, and reliability of the system. It also provides a way to integrate SMHP with utility grid in large scale, optimize electricity generation, and maximize the asset management.
- There are several potential site of MG in Nepal, at different stages of their evolution. The apex body of development and promoting of SMHP and MG is AEPC/RERL, which has recently installed and commissioned a pilot MG project at Kalungkhola corridor (Baglung district). The project has brought significant impact on the rural community, in social, technical, and economic

terms; thereby becoming source of research and experimental work for MG. RERL is also planning to interconnect it with NEA grid, for which series of dialogues and studies are underway

- Grid connection of the MG is a favorable solution from both MG and grid perspectives. Considering the fact that the MG can sell the surplus power at day time to the grid and draw additional power from grid during the period of deficit, grid connection is favorable to MG side. For NEA side, many households can be added as a customer without any investment. Technically, the connection of a power source at the endpoint of the feeder would help maintaining the system voltage and reducing the system losses.
- With the many opportunities with MG, there are some crucial challenges associated with it for speeding up the MG development in scale. Technically, it is a new technology, so the equipment and manpower need to be imported at present. Since the cost and the time associated with the installation, operation, and maintenance of imported equipment and technology is high, the technology should be developed at domestic level. Managerially, there is absence of suitable tariff, power exchange, and operational approach. Third, the investment portfolio is not attractive enough for business people; there is risk associated with the investment. These challenges draw due attention to multiple disciplines and necessitate rigorous dialogue. For this, a strong collaboration is needed between research institutes, government bodies, and private sector. RERL is expected to initiate and lead this role by making a formal body.
- Finally, a suitable guideline and framework need to development, which would reflect the entire context and directs the ways forward. Such guidance will provide information and required technical standards, managerial policy, and possible financial approaches for a given context.

Acknowledgments The author is thankful to Renewable Program for Rural Livelihood, AEPC for its support of this work. It has funded the work, provided valuable data and logistic support during the study.

Appendix

Figs. 9, 10, 11 and 12.

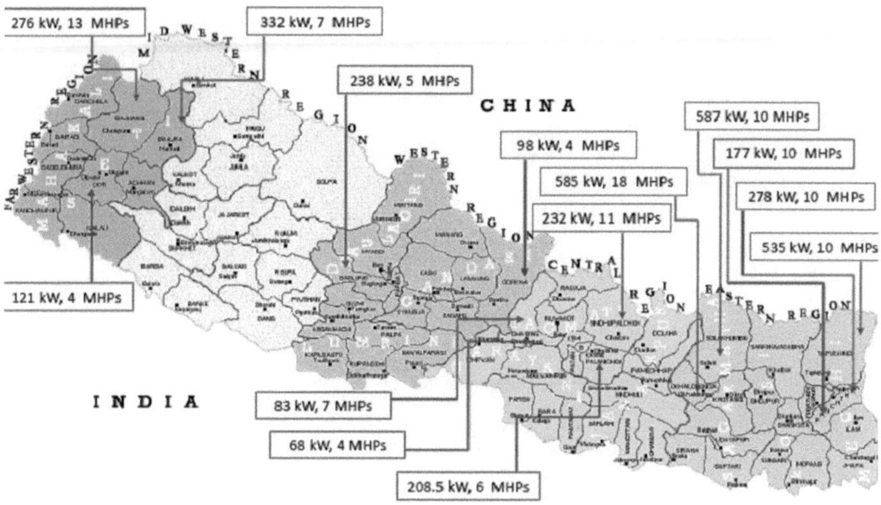

Fig. 9 Potential MG sites of Nepal [2]

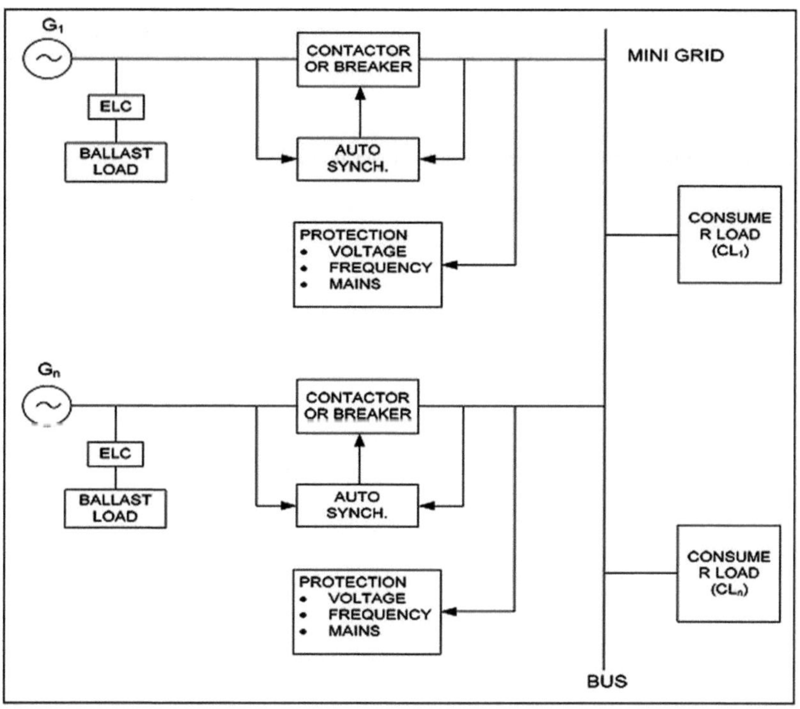

Fig. 10 Interconnection layout of SMHP-based mini grid

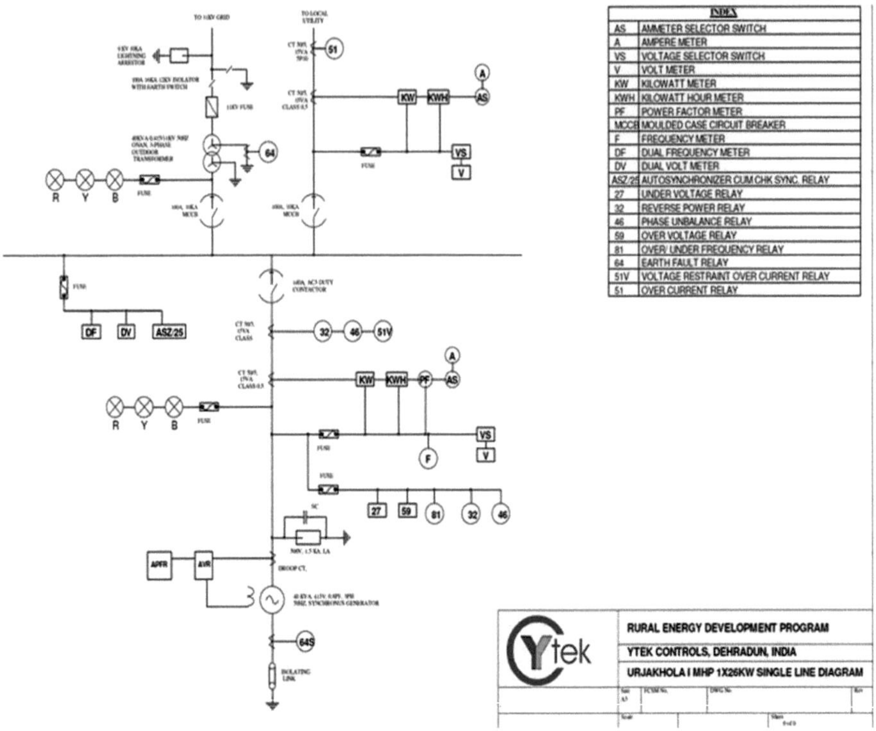

Fig. 11 Single line diagram of Theule Khola power house [3]

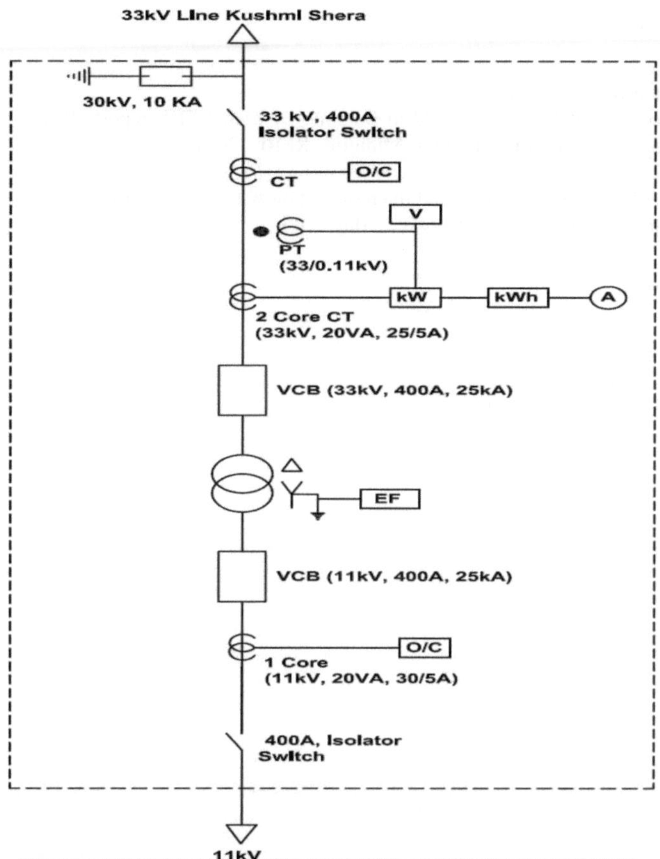

LEGEND

S.N	Symbol	Description
1	(transformer symbol)	Transformer
2	(isolator switch symbol)	Isolator Switch
3	□VCB	Vaccum Circuit Breaker
4	□ACB	Air Circuit Breaker
5	O/C	Over Current
6	EF	Earth Fault
7	KW	Kilo Watt
8	KWH	Kilo Watt Hour
9	V	Voltmeter
10	ROCOF/VVS	Rate Of Change Of Frequency/Voltage Vector Shift Relay
11	CT	Current Transformer
12	• PT	Potential Transformer
13	(lightning arrester symbol)	Lighting arrester

Fig. 12 SLD of interconnecting MG with grid at 33 kV feeder. In this case MG operates at 11 kV level, thereby desiring 33/11 kV interface for interconnection

References

1. NEA Annual Report, 2013
2. A Brief Note on piloting a mini grid in Baglung District, RERL report 2011
3. Detailed technical report on Baglung Minigrid, RERL, 2011
4. IEEE 1547 standards
5. Guidelines for power evacuation and interconnection with grid, Alternate Hydro Energy Center Indian Institute of Technology Roorkee, May 2011

MIX
Papier aus verantwortungsvollen Quellen
Paper from responsible sources
FSC® C105338

If you have any concerns about our products,
you can contact us on
ProductSafety@springernature.com

In case Publisher is established outside the EU,
the EU authorized representative is:
Springer Nature Customer Service Center GmbH
Europaplatz 3, 69115 Heidelberg, Germany

Printed by Libri Plureos GmbH
in Hamburg, Germany